Ice Age 2050s Certainty

by Rolf A. F. Witzsche

Contents

It takes an independent researcher to brake the taboos that have kept mainstream cosmology imprisoned, increasingly, during the past century, even while the universal taboos are known to be wrong.

For example, Johannes Kepler broke a science taboo in the early 1600s, that had kept astronomy imprisoned by a false concept for seventeen centuries. Epicycles and fudge factors had been invented to justify the taboos that stood contrary to physical evidence. One can see a lot of that in astrophysics today, where the evidence doesn't 'obey' the long-cherished assumptions that are nevertheless vigorously defended for numerous reasons, almost like a religious taboo. The Illustrated Science series is intended to open the scene beyond the threshold of the taboos, to where actual physical evidence speaks for itself.

The Ice Age and Climate Change sciences are riddled with obviously false assumptions that defy physical evidence. The evidence almost begs to be acknowledged. It promises amazing realizations when one begins to 'listen' to what the evidence is telling us, especially in astrophysics in connection with the coming Ice Age where so many myths abound that are simply not true, while the evidence tells us of the next New Ice Age beginning in the 2050s time frame, as truth, which has consequences that affect the entire world.

The Ice Age consequences as they are 'measured' in ice core data, render almost all areas outside the tropics as uninhabitable. This means that Canada, Europe, Russia, and parts of the USA, China, and India, need to be relocated into the tropics, for them to be able to exist past the 2050s time frame. The challenge to get this done is enormous, but it can be met if the scientific imperative for it becomes understood and acknowledged.

The scope of the existential challenge that the Ice Age brings with it, takes astrophysics out of the academic domain and places it into the foreground as one of the most critical issues of our time.

The big Climate Change events that have already worldwide effects are mere fringe effects in the flow of the ever-changing cosmic dynamics. The real, big effect, when the Ice Age begins, promises a dimmer and colder Sun with 70% less radiated energy. This defines our climate future.

Sure, we can live with all that by creating new platforms for agriculture that are able to operate under Ice Age conditions. But will we do it? The task is enormous. Will we fail ourselves on this front? We have no reason for that. We have the materials and energy resources on hand to accomplish everything that is required for us to continue to live. But will we do it? The big question, therefore, is; will we develop our inner resources as human beings sufficiently to get the job done, and to get it done in time? Or will we do nothing, ignore the challenge, and condemn our children and one-another to an agonizing death by starvation? That's the choice.

Towards meeting the inner challenge, I have created the epic series of novels, The Lodging for the Rose. And further, towards meeting the science challenge, I have produced numerous research books and several dozen exploration videos that the Illustrated Science series is modeled after. The work is the result of a quarter century of research, for which numerous elements of evidence in related fields came to light during this timeframe.

It is my hope that the work that went into these projects will help in some degree - for humanity that we are all a part of - to write itself a ticket to have a future.

About the book itself

Numerous fields of evidence tell us that the next Ice Age is near. Most of the evidence was discovered in the 1990s and thereafter. Some evidence is measured in ice cores, and some is measured in space by satellites. Some measurements are also made on the ground in terms of measurements of the Earth's magnetic pole drift observed in northern Canada. And all of this becomes combined with high-energy experiments at a leading national laboratory. Even the historic Little Ice Age, centered on the 1600s, takes on a new dimension in the light of this widely unfolding evidence.

Are ice ages real?

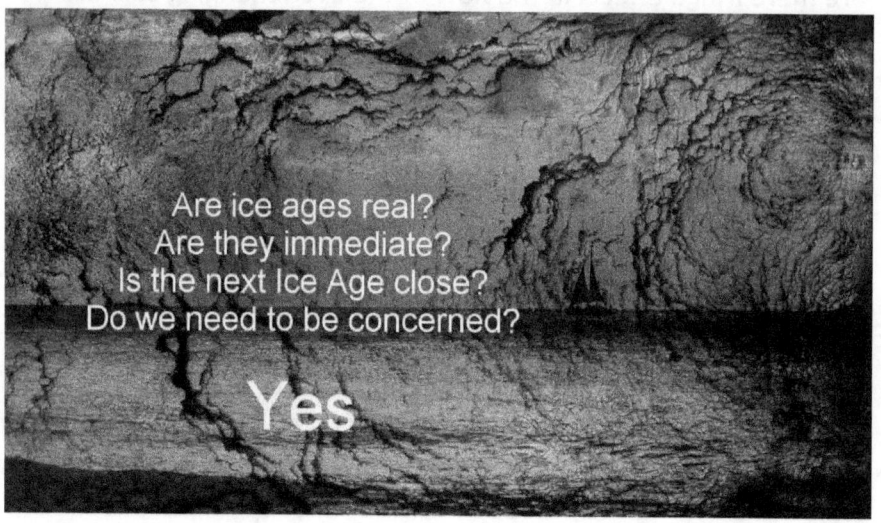

Are ice ages real? Yes!
Are they immediate? Yes!
Is the next Ice Age close? Yes!
Do we need to be concerned? Yes, again!

Friedrich Schiller, had lamented

*...all too often
the great moments in history
find society a little people.*

Friedrich von Schiller (1759 – 1805) German poet, philosopher, historian, and playwright

In the late 1700's Germany's great poet of freedom, Friedrich Schiller, had lamented that all too often the great moments in history find society a little people.

From Schiller's time to ours, more than 200 years have passed, and still the saying is true, perhaps more so than ever. Society has become increasingly small-minded, especially in the context of astrophysics that affects everyone's living on earth more deeply than anything else.

No one is not affected by the effects of the Sun

No one is not affected by the effects of the Sun on our planet. It becomes important therefore, to gain a wider understanding of the dynamics that affect the Sun, in order that we may adjust our living in accord with the changing dynamics.

When the Sun's radiated energy becomes reduced

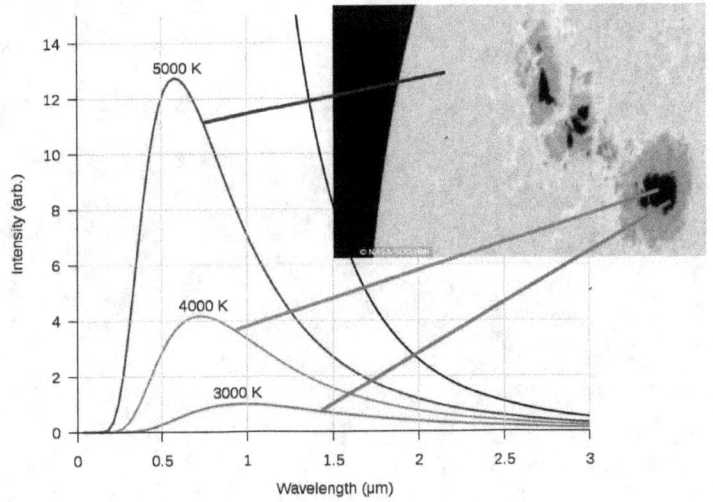

Our understanding of the cosmic dynamics is especially critical in the case of the New Ice Age before us that promises to have gigantic consequences on Earth when the Sun's radiated energy becomes reduced by 70% on the day the Sun goes inactive. That's the day when its external plasma-fusion process falls below its minimal supply threshold and stops. The Sun then becomes as dim as we see it below the sunspots.

Potentially as near as the 2050s

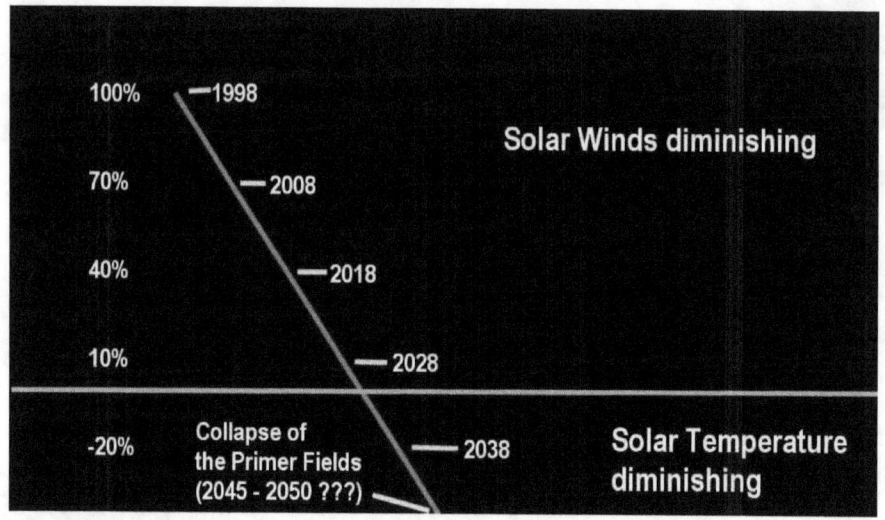

The day for this transition, by which the Ice Age begins, is near in time, potentially as near as the 2050s. That's when the energy input drops below the minimal threshold that is required for the Sun to remain active. The solar-wind pressure is a perfect indicator of the still remaining strength and its rate of diminishing. When it drops to zero, the fusion process is doomed to suffer a chain-reaction collapse.

The liveable zone on the Earth shrinks

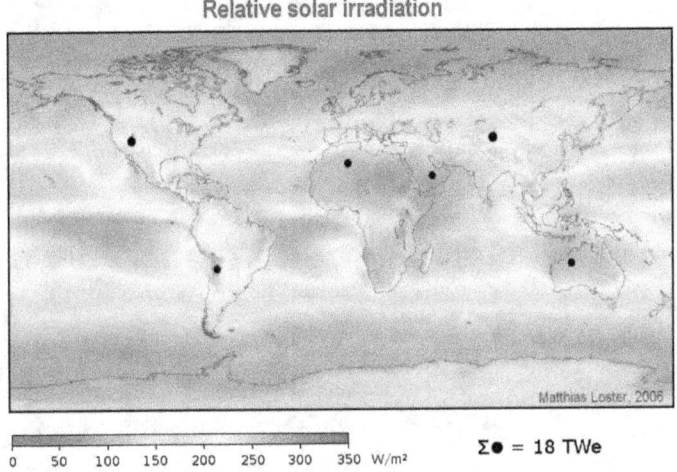

Relative solar irradiation

$\Sigma\bullet = 18$ TWe

At this point the liveable zone on the Earth shrinks to the narrow band of the tropics where the solar exposure is the highest, so that 30% of the current solar radiation will be sufficient to sustain life. While we still have 30 years remaining to adjust our living in accord with the shifting cosmic dynamics, such as by building the infrastructures in the tropics that will be needed at this point, for relocating all the northern nations into the tropics, in real terms we are running out of time already.

Vast infrastructures will have to be in place

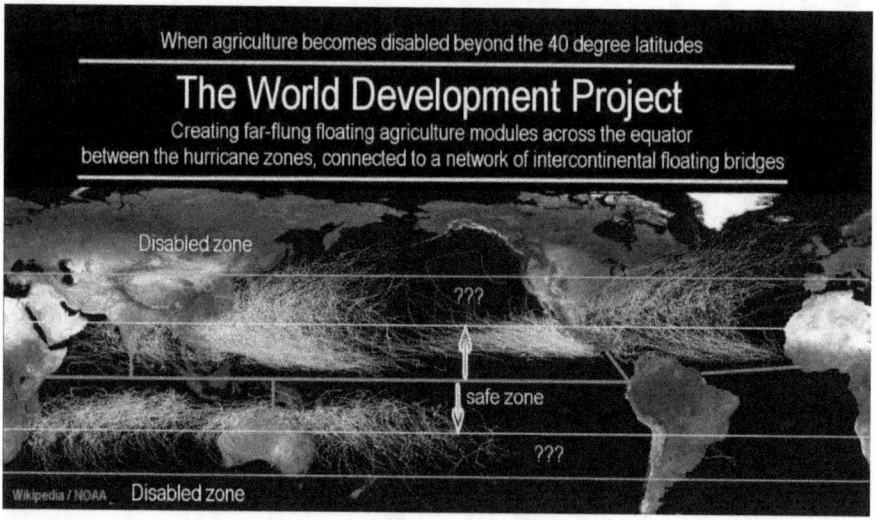

The vast infrastructures will have to be in place and be fully operational before the northern territories become uninhabitable as the result of the 70% reduction of the Sun's radiated energy, because at the point of this transition the liveable area of the world will shrink from the day on when the Sun goes inactive and the Next Ice Age unfolds.

The consequences of an inactive Sun

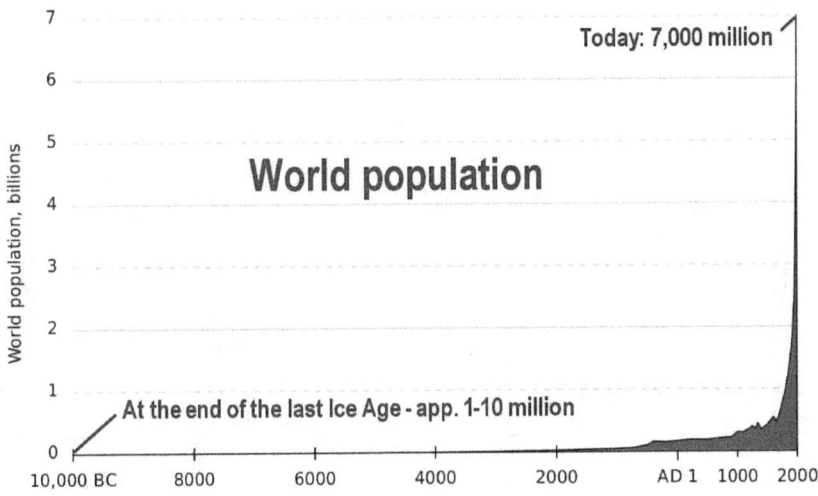

The consequences of living under an inactive Sun are so immense that humanity as a whole came out of the Last Ice Age with a tiny world population of a mere 1-10 million people.

Yes, 1-10 million people

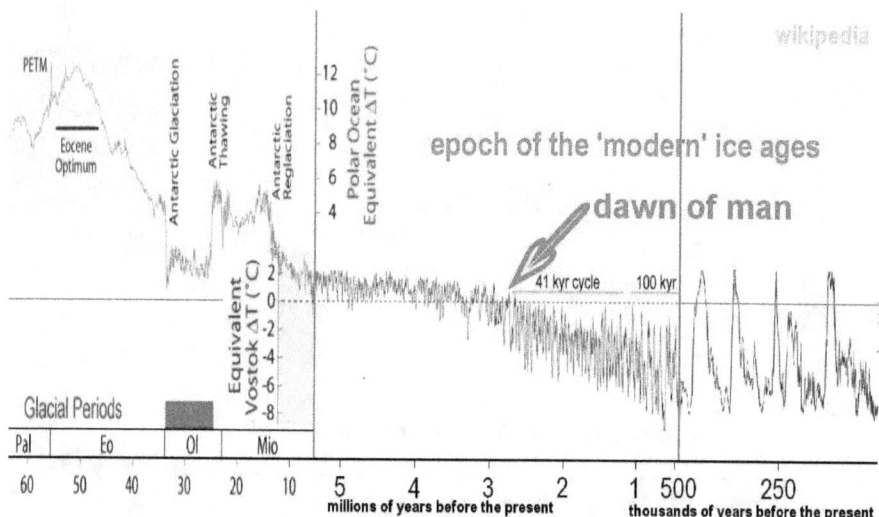

Yes, 1-10 million people, that's all we had left after 2 million years of human development, across more than 20 ice ages.

The depressing effect of the Ice Age

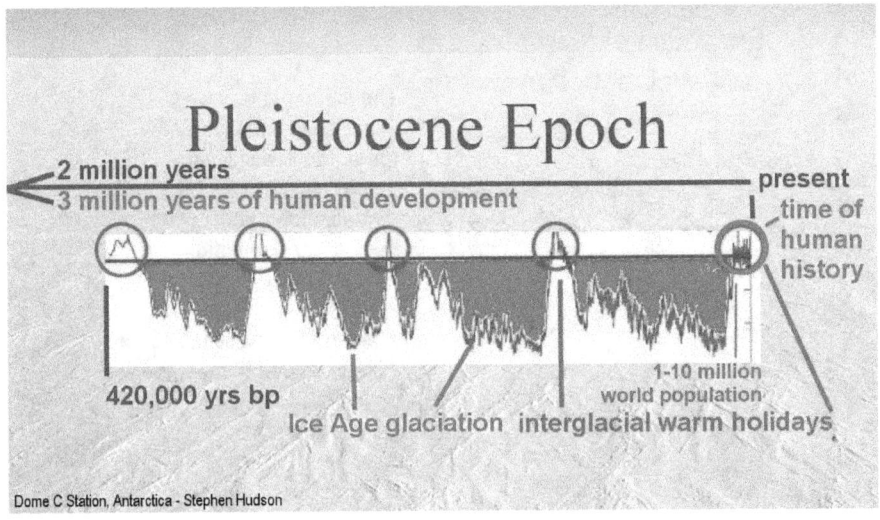

Dome C Station, Antarctica - Stephen Hudson

The depressing effect of the Ice Age environments on human living has evidently been so immense that the concept of human history is typically limited to only the minuscule time of the 12,000 years of the current interglacial period, as if nothing had existed before.

All the creation sagas of the great religions, of the origin of man, are located in the brief period of the current interglacial, which is itself an anomaly.

On the long timeframe of the Pleistocene Ice Age Epoch, the interglacial warm periods, especially those of the last half million years, amount to only 15% of the time, with deep glaciation occurring for the remaining 85% of the time, which renders the glacial environment the normal environment of the Earth during the Pleistocene Epoch.

The ocean levels had dropped 400 feet

Expect a near-term sea-level reduction

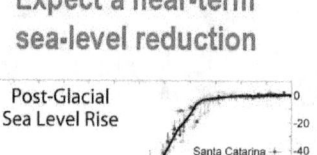

Post-Glacial Sea Level Rise

Meltwater Pulse 1A

Last Glacial Maximum

Santa Catarina
Rio de Janiero
Senegal
Malacca Straits
upper bound
Australia
Jamaica
Tahiti
Huon Peninsula
Barbados
lower bound
Sunda/Vietnam Shelf

Sea Level Change (m)

0
-20
-40
-60
-80
-100
-120
-140

24 22 20 18 16 14 12 10 8 6 4 2 0
Thousands of Years Ago

The danger is that we will experience a massive reduction in sea level in the near term as the re-glaciation begins with the Ice Age Transition now in progress.
Picture a loss of only 20 meters A powerfull new renaissance will be needed to meet the physical challenge for infrastructures

During the last glaciation environment water vapor had turned into snow and had accumulated on land so massively over time that the ocean levels had dropped 400 feet.
The vast historically known sea-level reduction amounts to ice accumulations on land on a scale that defies the imagination. The ice has shaped the landscape.

The great inland sea in northern Canada

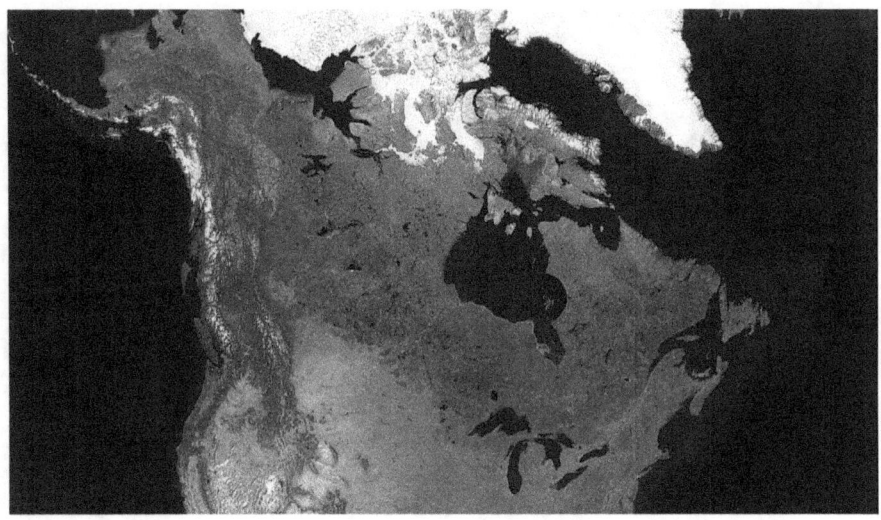

The great inland sea in northern Canada, named Hudson Bay, is believed to have been formed as a depression of the crust of the Earth by the weight of ice accumulations. The great lakes in southern Canada, may have been carved by moving ice masses. That's the kind of effect that an inactive Sun has had on our planet.

When the Sun goes inactive

Ice Age of the dimming Sun in 30 years

That's the kind of climate future that available evidence suggests will be upon us again, starting in roughly 30 years when the Sun goes inactive. That's what we need to prepare ourselves for, in advance of its happening.

Relocating ourselves into the tropics

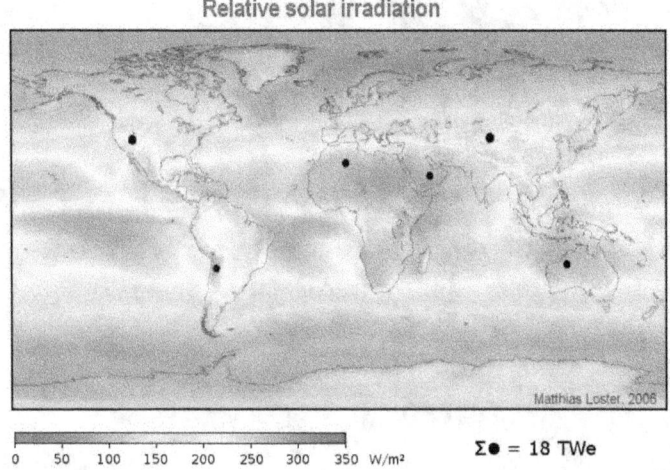

Relative solar irradiation

$\Sigma \bullet = 18$ TWe

While we can continue to live in the energy-reduced climate by relocating ourselves into the tropics, no intention towards addressing this challenge is evident, or it to be even considered. Instead, society keeps its mind and eyes closed.

Friedrich Schiller's lament still applies

...all too often
the great moments in history
find society a little people.

Friedrich von Schiller (1759 – 1805) German poet, philosopher, historian, and playwright

And so, once again, Friedrich Schiller's lament still applies, that all too often the great moments in history find society a little people.

Almost all of the great pioneers in the world

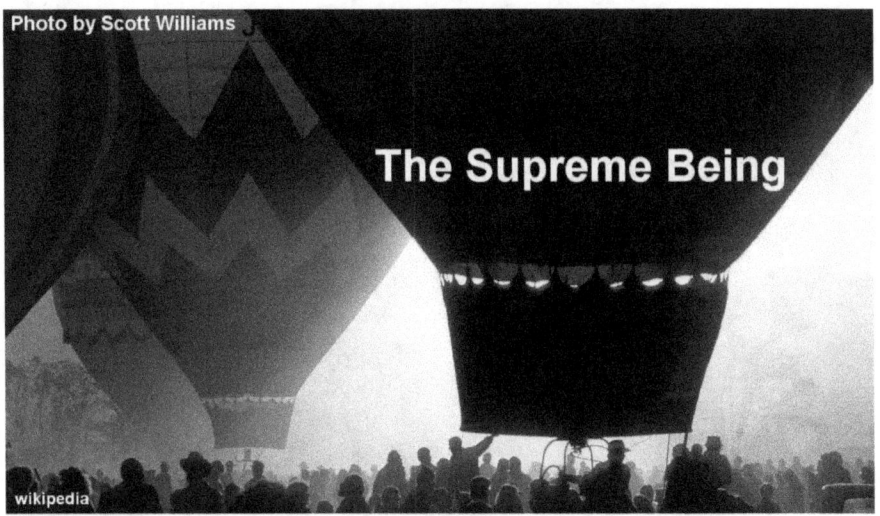

Photo by Scott Williams

The Supreme Being

wikipedia

Even the world's greatest champions for scientific progress on the entire front continue nevertheless to live with their mind and eyes closed today on the most critical issues that they do not wish to see, even in the face of monumental evidence that the universe and humanity with it, is vastly richer and more powerful than they dare to imagine.

Almost all of the great pioneers in the world who sing the tune of science vigorously, refuse nevertheless to acknowledge that the human being has the capacity with the mind to recognize the principles that operate in the universe, and to look with this mind forward into the future to behold the unfolding of dynamic effects on the cosmic scale 30 years before the visible consequences come to pass.

Only the human being has this kind of capacity on the Earth, but as Friedrich Schiller laments, society, in spite of its great advances, still refuses all too often to acknowledge itself as human beings. It lets the critical opportunities that open doors to great freedoms and unsurpassed creations, pass by unrealized. That's a tragedy that can be healed.

Time remaining to meet the Ice Age challenge fully

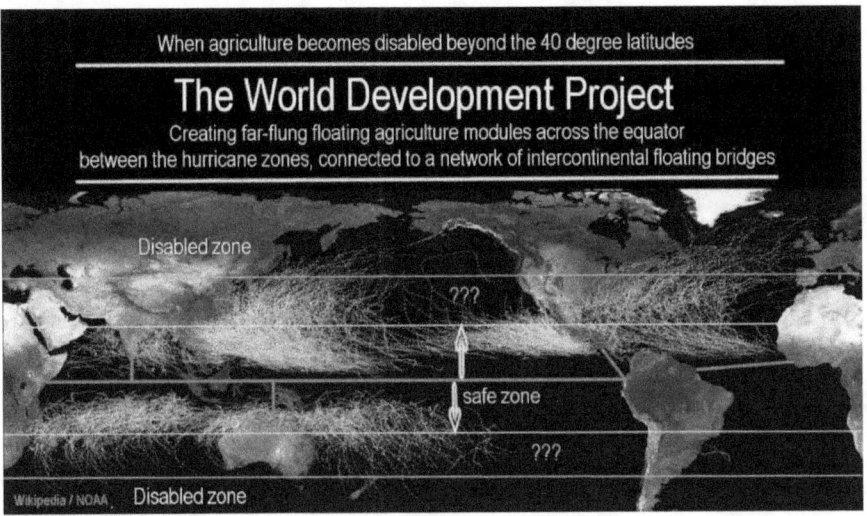

We still have time remaining to us to meet the Ice Age challenge fully, to place our agriculture into the tropics, afloat on the seas for the lack of land in the tropics, together with the building of 6000 new cities for a million people each, complete with new industries, all connected with floating bridges and high-speed railways. Any lesser response amounts to society committing collective suicide. Fortunately, it is physically possible to build all of the required infrastructures in the 30 years we have remaining in the current interglacial period, but we won't do it, if we don't get started. What hinders us in the present is the devil that is in the mind. That is what needs to be challenged. The devil is small-minded thinking, a lack of universal love for our humanity. With the mind blinded by small-mindedness, society has placed itself into a box that has become a trap. This trap is real. It has consequences, especially in astrophysical science.

The box of perception in astrophysics

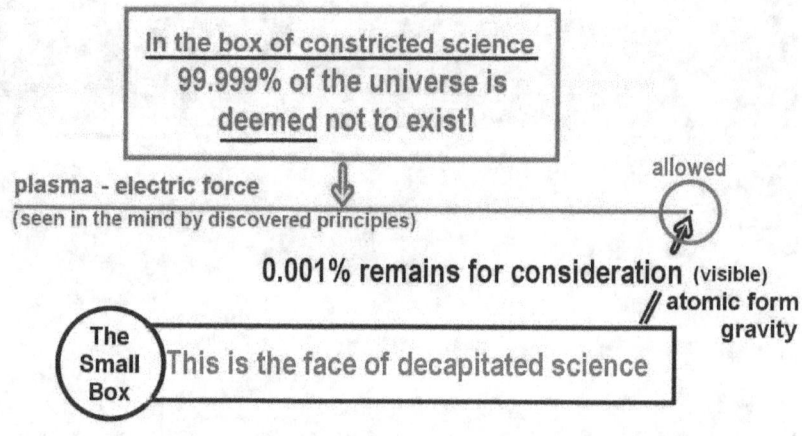

In the box of constricted science
99.999% of the universe is
deemed not to exist!

plasma - electric force
(seen in the mind by discovered principles)

allowed

0.001% remains for consideration (visible)

// atomic form
gravity

The Small Box

This is the face of decapitated science

The box of perception in astrophysics has been kept intentionally so small that 99.999% of the mass of the universe is deemed not to exist. The keeper of the box commands: What you see is what you get. You see the consequences as they happen. You do not see principles that cause them. Universal principles do not exist. Plasma in space that you cannot see with the eyes, does not exist, even when the effects are visible. You are allowed to recognize gravity, and you are commanded to recognize it as the only force of the universe. The 'command' is, that anything else does not exist.

Of course, the constricted box in which 99.999% of the mass of the universe is deemed not to exist, is rather empty, isn't it? In this box less than a thousands of a percent of the riches of the universe is allowed to be recognized. The keeper of the box tells society, don't you dare poke a hole through the box and look at the real world. The command reverberates: It doesn't exist.

Nevertheless, it does exist. The reality is, that the dream-images that are conjured up in the prison of the box, are necessarily fairy tales.

Plato illustrated the box as a prison

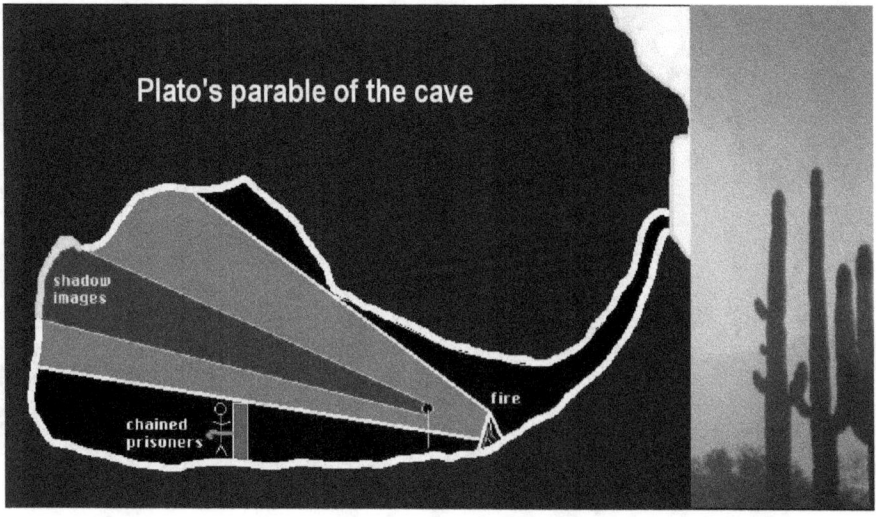

Plato illustrated the box as a prison in the form of a cave inside a mountain. The prisoners are chained behind a wall. All they ever see are reflections on the wall of a fire that is operated by the keepers of the prison, and some shadow images produced by the keepers. But one day, one prisoner brakes the chains, climbs over the wall, and instantly recognizes that the shadows that they all saw, which defined their world, was an illusion. In time the prisoner notices the exit from the cave, and finds a great wide world outside that he had never imagined to exist.

When Schiller laments that society too often fails to respond to the great opportunities before it, he laments in a sense that too much of society remains imprisoned by its small-minded, captivating, and often intentionally cultivated, illusions.

Obviously, if one is psychologically imprisoned into a science that disregards 99.999% of the universe as if it doesn't exist, one lives in a world of fairy tales where extremely little is actually real.

Imprisoned for a thousand years by the doctrine of the church

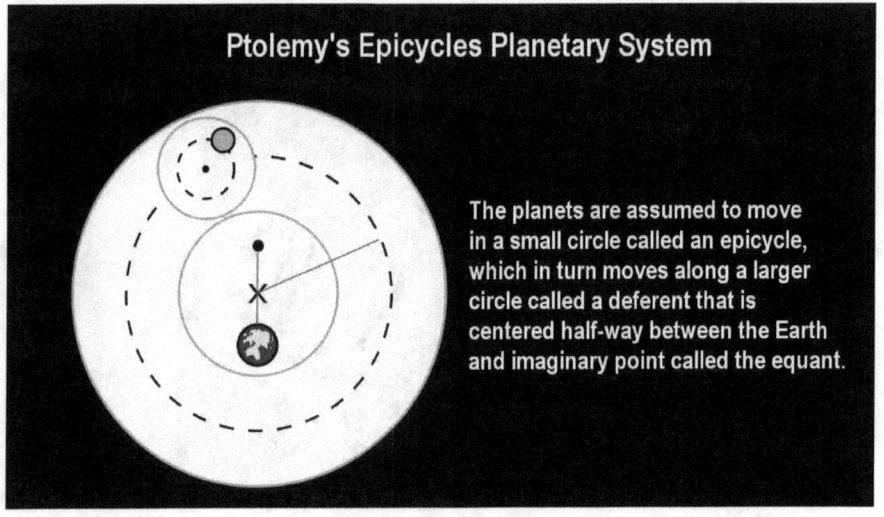

Astronomy had been imprisoned for a thousand years by the doctrine of the church that all orbits in the heavens must follow perfect circles, because the circle is a perfect geometric form, consequently the heavens cannot contain anything less.

The astronomer Johannes Kepler

Johannes Kepler
German Astronomer
1571 - 1630

Discoverer of the
universal force of gravity
and it principles manifest
in planetary motions

The astronomer Johannes Kepler was the first man in history who climbed out of this cave and discerned in the mind the greater reality that the senses can never actually see. He liberated astronomy with his discoveries of principles that can only be seen in the mind, by recognizing their effects in the visible universe.

Kepler had discovered the result of a process

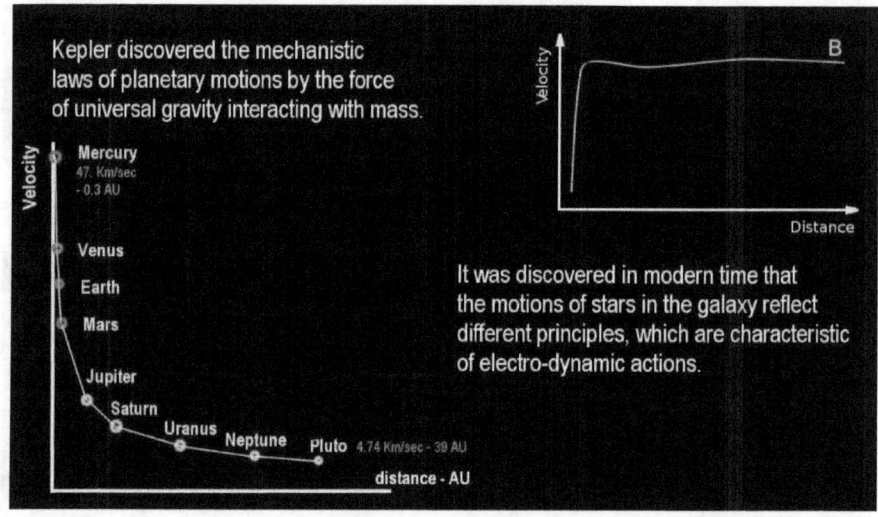

Kepler had discovered the result of a process and the principles expressed by it. However, he did not discover the forces that created the process and maintain it. The required technological capability for this advanced discovery to be possible didn't exist in his time, but it exists now.

Nevertheless, we are facing the same type of challenge today that he faced, to see with the mind, such as the recognition of plasma in cosmic space and its dynamics, which too, can only be recognized by their effects in the visible universe. And this we are capable of. We can see with the power of the mind the great electromagnetic forces of the universe in action, which are inherent in plasma in space, but which no eye has ever seen, or ever will be able to see. We can only physically see the consequences of the great cosmic forces. Those consequences are actually amazingly massive, as the story of the ice ages hints at.

The year 1990 was the pivotal year

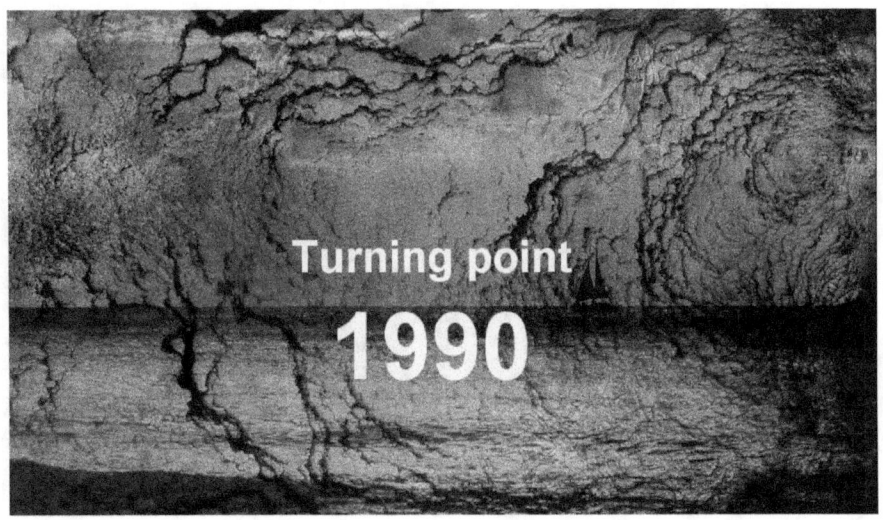

Most of the hard evidence that we have of the previous ice ages, is relatively recent. Almost all of it didn't exist prior to 1990. The great research efforts and resulting discoveries began around 1990.
The year 1990 was the pivotal year when the scientific ventures related to the ice ages were beginning to bear fruit. The big ice core drilling projects, both in Antarctica and in Greenland, were launched in the 1990s, and in some cases slightly before.

The Ulysses spacecraft was launched

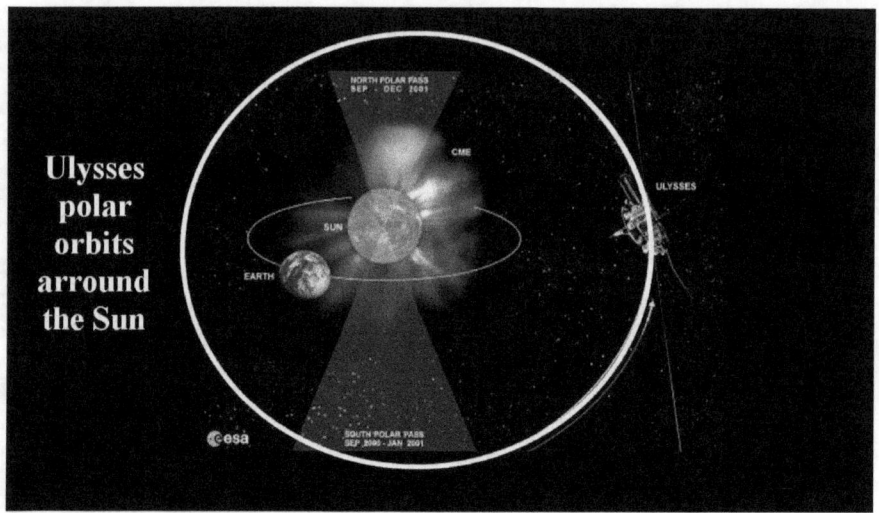

The year 1990 was also the year when the Ulysses spacecraft was launched into a polar orbit around the Sun that gave us the cleanest and most direct data of the dynamics of the Sun, especially the solar-wind patterns.

Ulysses gave us a clean look at the solar-wind pressure

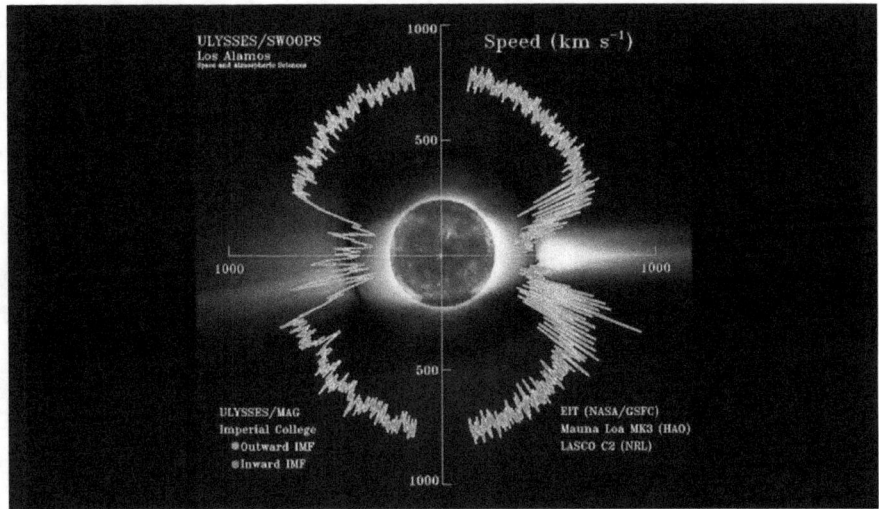

From its wide polar orbit around the Sun, the Ulysses instrumentation was able to measure the characteristics of the solar wind without the strong interference by the heliospheric current sheet that extends from the equator of the Sun to the edge of the heliosphere. Ulysses was the first and only satellite we ever had that gave us a clean look at the solar-wind pressure, which is a critical component for understanding the solar factor of the Ice Age dynamics.

From the 1990s onward

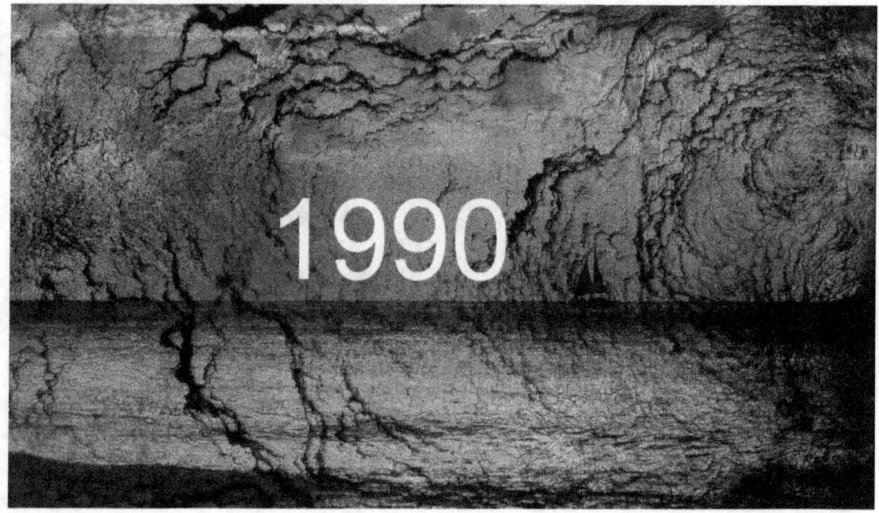

All of this means that humanity's ice age science hadn't had any hard and fast physical data to work with prior to 1990 when the big ice-drilling projects began and the Ulysses satellite was launched. The development of this data, from the 1990s onward, renders the Ice Age science a fairly new discipline, as all prior assumptions were built on just vague evidence or mere conjecture.

Essentially incomplete till the early 2010 timeframe

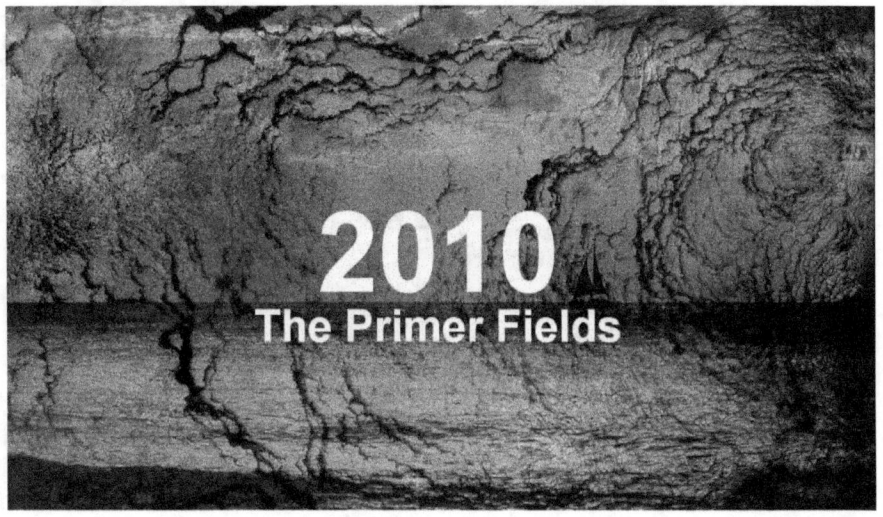

And even this advance was essentially incomplete till the early 2010 timeframe when the breakthrough discoveries were made, of the primer fields for the electric Sun, for which verification exists in the Ulysses data, which will be shown later. This means that our understanding of Ice Age physics and its consequences is far from being an old-hat issue.

The very concept of an ice age is relatively new

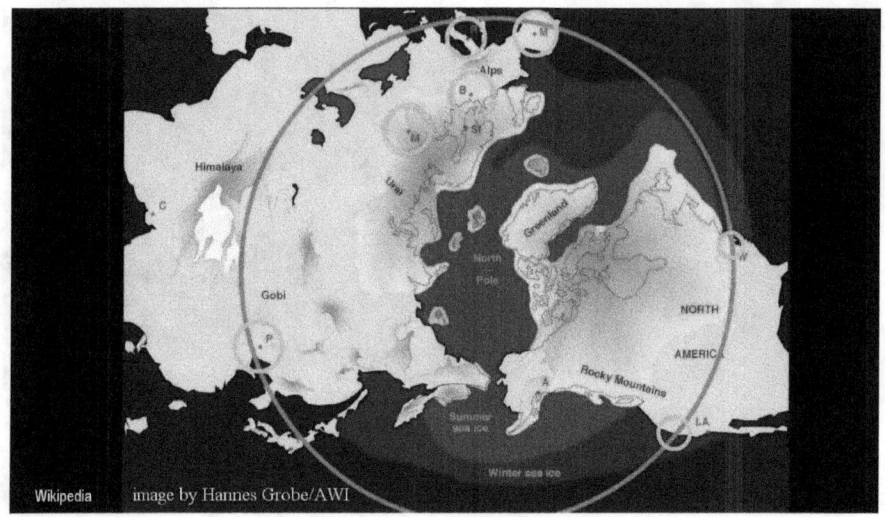

In fact, the very concept of an ice age is relatively new. The occurrence of vast global ice ages in the history of the world hadn't even been postulated until the 1800s. The concept simply didn't exist prior to the 1800s.

The Little Ice Age that gripped Europe

Hendrick Avercamp

The Little Ice Age that gripped Europe, centered in time on the mid 1600s, had been seen as nothing more than just a strange climate anomaly, when it occurred. To some degree it is still so regarded by many people who are living in the boxed-in world of constricted science.

Way back in time, in the late 1500s, during the early stages of the Little Ice Age, the world had suddenly become colder. No one knew why. Rivers and canals became skating rinks in the winter as this painting from 1608 illustrates. Later, when the cooling deepened, farms and villages in the Swiss Alps were destroyed by encroaching glaciers. This occurred some time during the mid-seventeenth Century.

Even the River Thames froze over

The Frozen Thames, 1677 - Abraham Hondius - wikipedia

Even the River Thames froze over at this time, as the painting shows from 1677. The summers too, were cold and wet, and with great variability. Crop practices had to be altered to adapt to the shortened, less reliable growing season. Many years of dearth and famine resulted from this climate change.

The famines were less popular

The Reverend Robert Walker
Skating on Duddingston Loch,
attributed to Henry Raeburn, 1790s

At the time, ice skating became rather popular as this painting from 1790 shows. The famines were less popular, though. Famines in France and Sweden claimed roughly 10 percent of the population, and in Estonia 20%, and in Finland more than 30%. These were hard times.

The cold period lasted into the mid 1800s

On the main canal of Pompenburg, Rotterdam in 1825, by Bartholomeus Johannes van Hove

The cold period lasted into the mid 1800s, as the scene shows from Rotterdam in 1825. Nor had the Little Ice Age been a local phenomenon, as some scientists still claim.

Ice core records from Antarctica

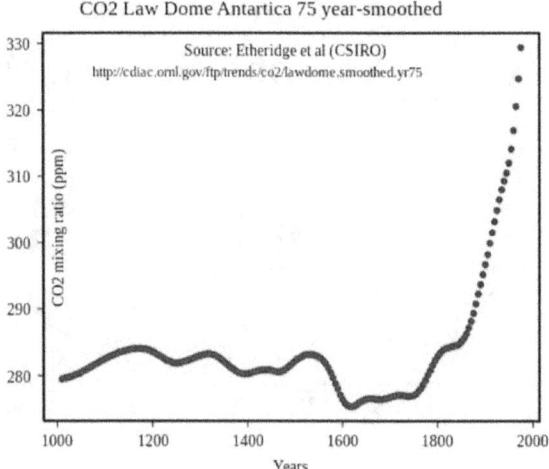

Ice core records from Antarctica show a marked drop in CO2 concentration in the ice, for the period of the Little Ice Age. Lower CO2 levels typically reflect colder global temperatures.

Coincidence of the Maunder Minimum

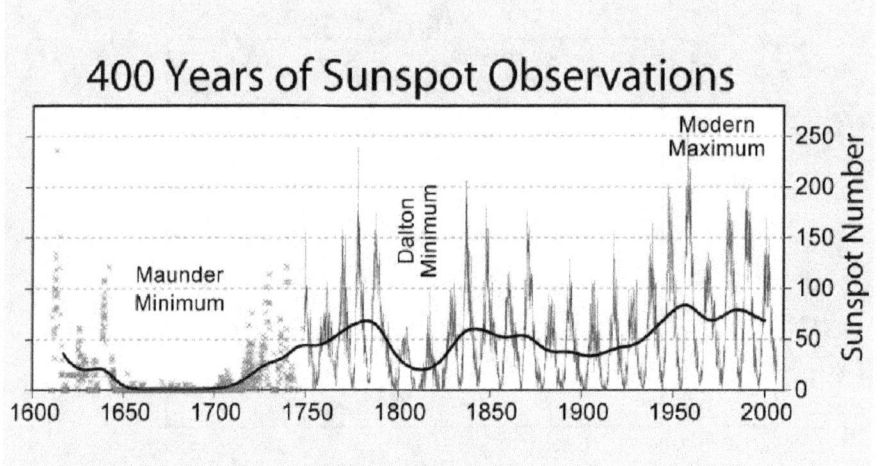

It was discovered in much later times, that in the middle of the exceedingly cold period, almost no sunspots had occurred on the face of the sun. The lack of sunspots became known as the Maunder Minimum. It extended from 1645 to 1715.

For a single stretch of 30 years within this period, fewer than 50 sunspots were observed, as compared with 40,000–50,000 spots in modern times. However, the coincidence of the Maunder Minimum with the cold period hadn't been recognized until the late 1800s. It had been regarded as unimportant earlier, just as the Ice Age phenomenon itself had been regarded as unimportant then. The recognition was beginning to dawn much later in the 1800s.

No one at the time imagined that large-scale climate changes

Chamonix valley 2010

A geographer, who reported on his journey through the Alps in the mid 1700s, had noted large boulders in the valley that shouldn't be where they were. He was told by the locals that the glaciers had once extended far down into the valley, which had probably brought the boulders with them. With the 'discovery' of other such boulders in strange places, near glaciers, the idea began to dawn that major glaciation periods had occurred in the past.

With it, as an explanation, the idea emerged in the mid 1800s that the climate changes for the theorized ice ages, must have been caused by changes in the Earth orbit. No one at the time imagined that large-scale climate changes could be caused by factors that affect our Sun directly. The Maunder minimum hadn't been seen as related evidence then. This connection hadn't been recognized at the time, or it had not been known at the time.

The connection between changes in solar activity and the climate on Earth, is increasingly being acknowledged. This means that it is also being acknowledged in leading-edge circles that the Sun plays a causative role in climate dynamics.

While the causative role of the Sun is still under dispute by some, the theory that orbital cycles must cause the ice ages is likewise still upheld by them, though no real evidence exists in support of the theory.

The modern version of the orbital cycles theory

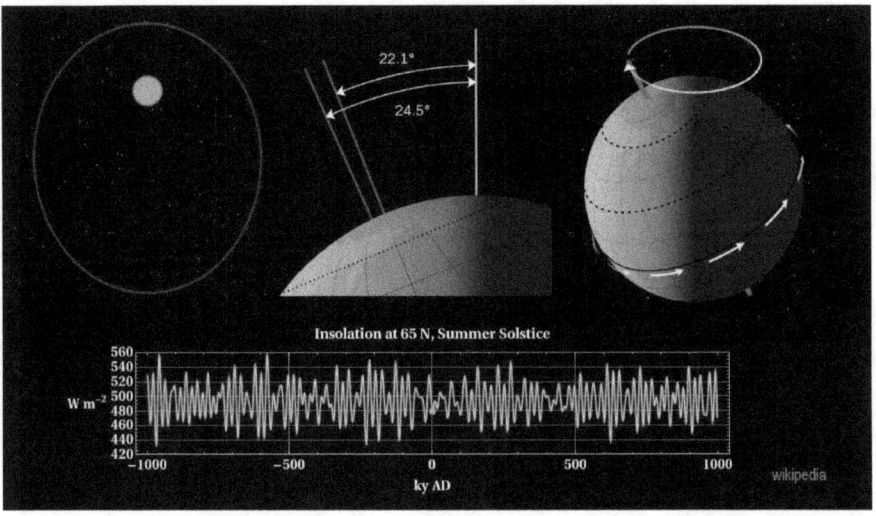

The modern version of the orbital cycles theory is the Milankovitch cycles theory. It combines three minuscule cyclical variations, spanning tens of thousands of years each, into a composite climate effect. The theory breaks down, however when one considers that every single one of the considered cyclical variations has no net effect on the total sunlight received on Earth, even by changes in the orbital eccentricity, as Johannes Kepler had discovered already in the 1600s.

The orbital cycles only alter the seasonal and hemispheric distribution of the solar radiation received, which are too minuscule to cause the gigantic Ice Age phenomena.

The evidence that the Ice Ages are global

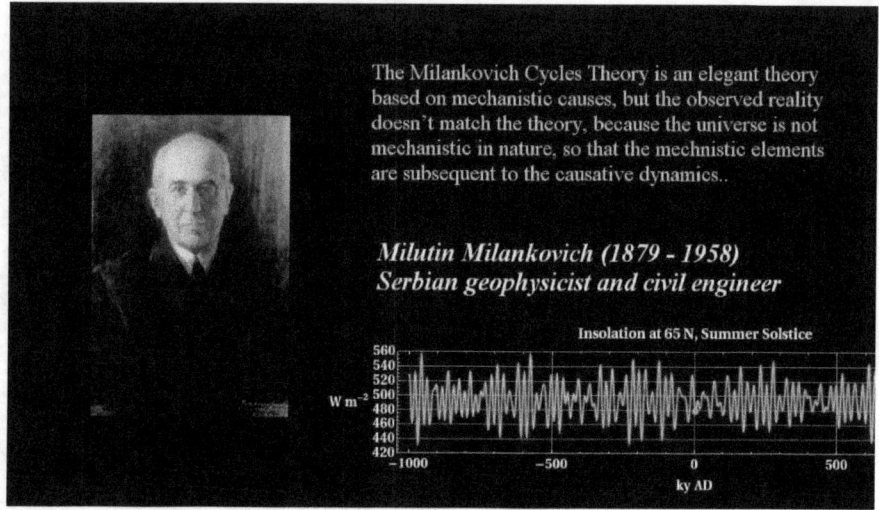

The Milankovich Cycles Theory is an elegant theory based on mechanistic causes, but the observed reality doesn't match the theory, because the universe is not mechanistic in nature, so that the mechnistic elements are subsequent to the causative dynamics..

Milutin Milankovich (1879 - 1958)
Serbian geophysicist and civil engineer

Insolation at 65 N, Summer Solstice

The evidence that the Ice Ages are global, and not hemispheric, did not exist at the time the Milankovitch theory was created, which might not have been created had this fact been known at the time.

Evidence for the global ice age is extremely new

Vostok Station in Antarctica

The evidence for the global ice age is extremely new, so that concrete evidence didn't exist until the early 1990s when the deep ice cores were drilled both in Greenland and in Antarctica, which both presented identical results.

The Russian station, Vostok

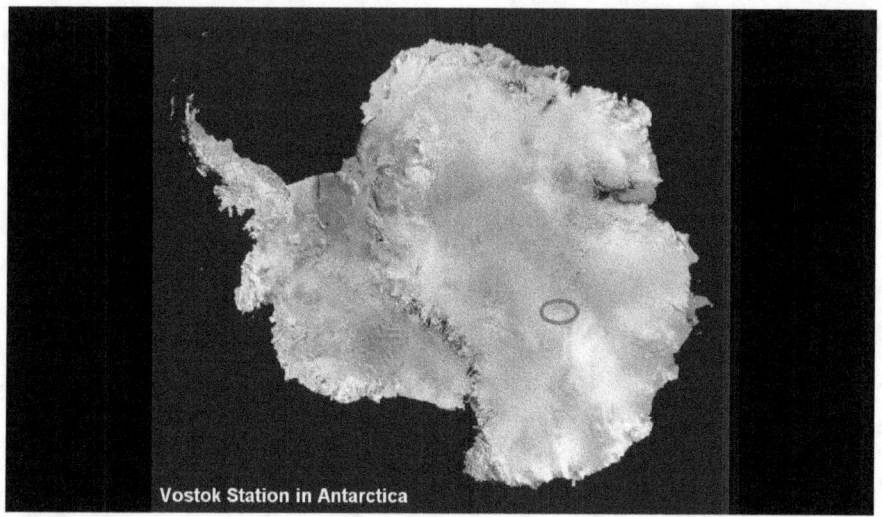

Vostok Station in Antarctica

The Russian station, Vostok, was set up in 1957, and remained manned year-round thereafter at the coldest spot on Earth, measured at -89.2 C.

Vostok had drilled its first deep hole in 1990

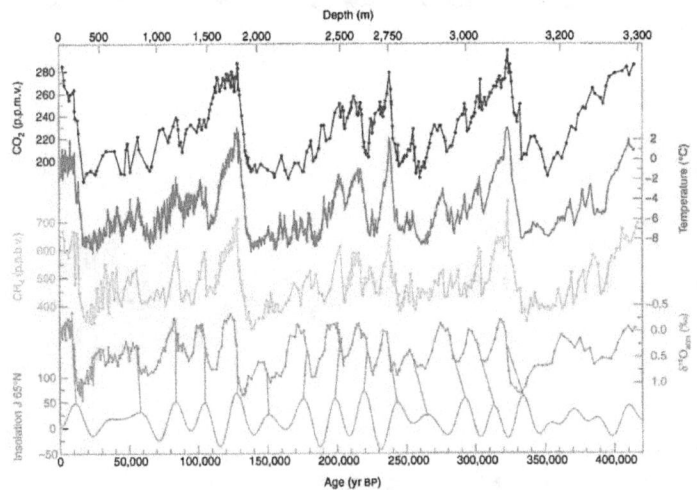

Vostok had drilled its first deep hole in 1990, 2,200 meters deep into the ice. This was a big achievement. The retrieved samples cover a period of 150,000 years into the past. Later drilling reached a depth of 3623 meters in 1996, which cut through 420,000 years of ice accumulation, spanning almost 4 full glaciation cycles.

The Greenland Ice Core Project

The Greenland Ice Core Project, in central Greenland, completed its first deep drilling in 1992.
The project had drilled 3028 meters deep into the ice.

Ice from the last 100,000 years

It had extracted ice from the last 100,000 years, containing records of the climate in the northern hemisphere.

Proof that Ice Ages are a global phenomenon

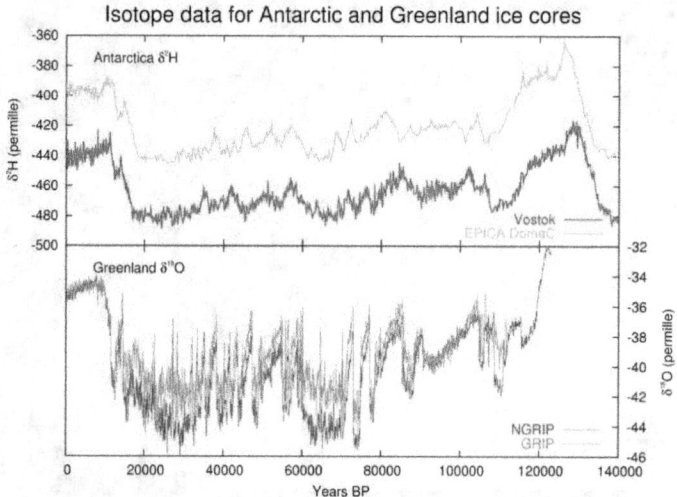

Isotope data for Antarctic and Greenland ice cores

The results obtained from Greenland, as one would expect,
matched closely the results obtained from Antarctica.
The coincidence in the ice from these widely separated drilling sites,
provides solid proof that Ice Ages are a global phenomenon.

Measurements of deep ocean sediments

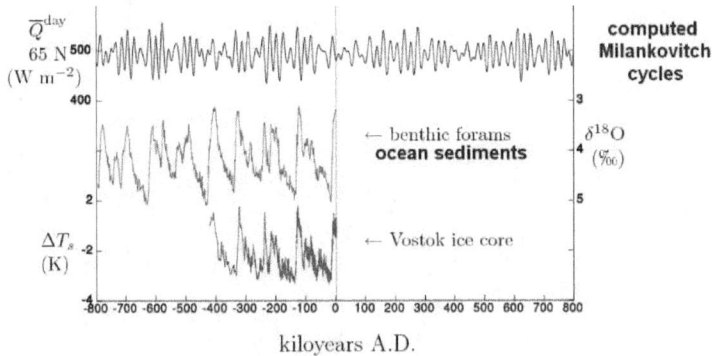

comparison of measurements between
ice core data and ocean sediments

The ice cores samples obtained at the Vostok site in Antarctica,
further agree with measurements of deep ocean sediments.
Of course, all these measurements themselves do not indicate, or
even proof, what had actually caused the ice ages. Or do they?

The Greenland ice cores tell us

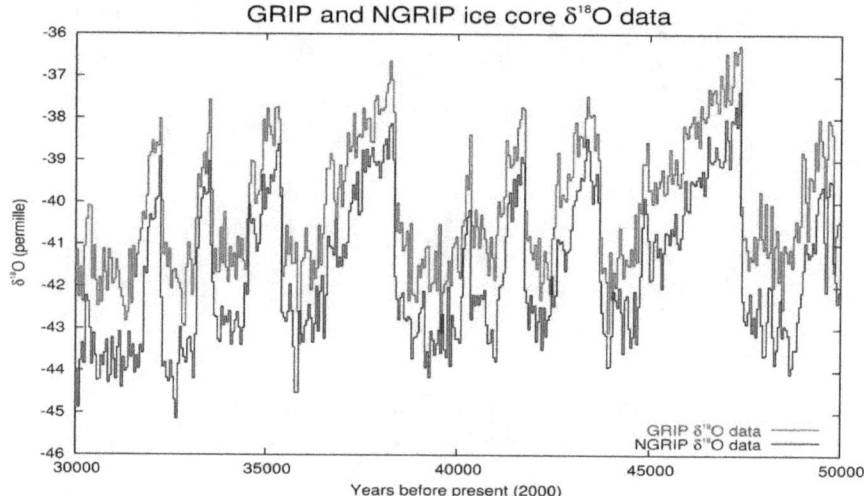

The Greenland ice cores tell us that the Ice Age itself, which cuts very deep into the negative temperatures, has many enormous fluctuations occurring within them. Some are reaching from the deepest low-level, almost instantly, to near the interglacial level, with transition times sometimes measured in just decades.

The NGRIP project

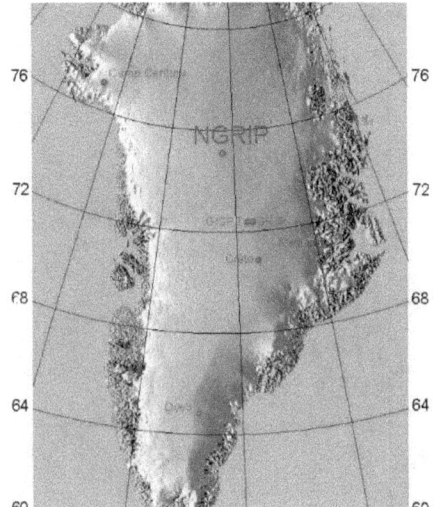

That these immense temperature swings during the last Ice Age are not anomalies of a specific drilling site on Greenland, became apparent when the deep drilling was repeated a decade later at another location on the Greenland Ice Sheet, further north, called the NGRIP project.

Large fluctuations do not show up in Antarctica

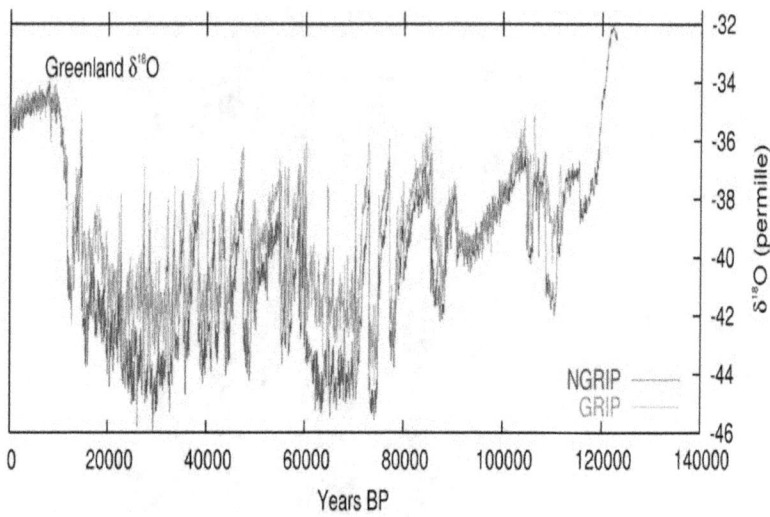

Essentially the same fluctuations were measured there, as were measured at the more southern site. The fact that the large fluctuations do not show up in Antarctica, may be due to the fact that Antarctica is essentially an ice desert that gets not enough accumulation to preserve the fine details.

It is evidently for this reason, too, that the Antarctic ice is sampled for changing ratios of deuterium in the ice, which is a heavy isotope of hydrogen, while in the 'wetter' Greenland the ratio of the heavy isotope of oxygen, named oxigen-18, is measured. These differences too, affect the details that can be observed.

Details in the Greenland ice 'speak' of immense events

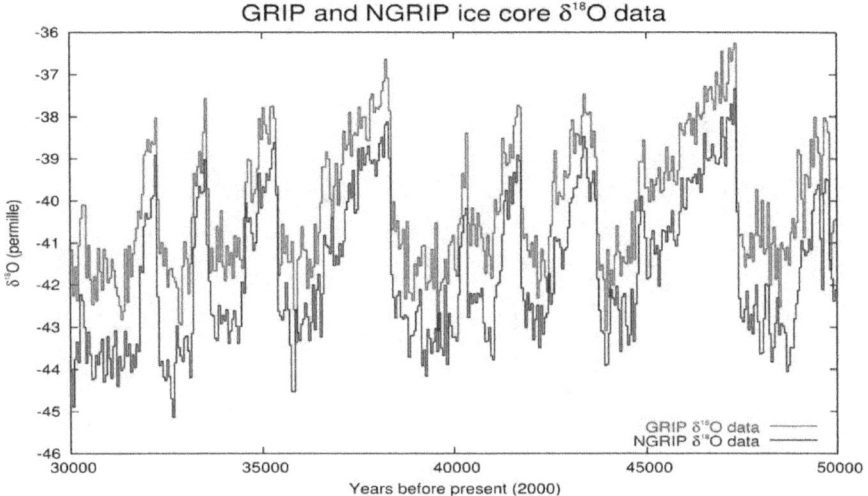

All of this tells us that the sharply defined details in the Greenland ice are real, and that they 'speak' of immense events that have occurred during the last glaciation period.

The truly immense events evidently reflect immense causes, such as the Sun becoming re-activated briefly at regular intervals during the glaciation period. Attempts have been made to explain the gigantic events as resulting from ocean current fluctuations. That's hardly a reasonable possibility in a largely frozen Ice Age world.

Look at the Little Ice Age in comparison

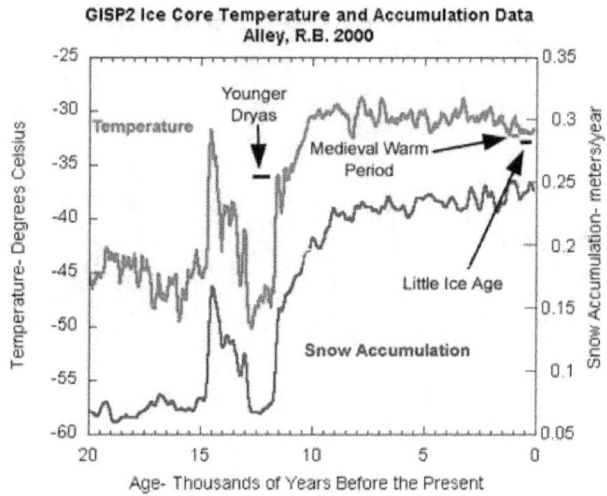

Now look at the Little Ice Age in comparison, which we do have records of. The Little Ice Age represents a half a degree of cooling on the cosmic scale. Sure, on the ground the effect seemed big. It changed the landscape and caused famines. But in real terms, it is almost unrecognizable on the larger scale where we see fluctuations happening that are 40 times as extensive.
Immensely huge events have cause these gigantic fluctuations. The cause for these enormous effects can only be attributed to the Sun. The Maunder Minimum points in this direction. During the 'tiny' Little Ice Age the Sun had weakened enough to loose its sunspots. For the huge events, we will likely see the Sun going inactive completely, because nothing else makes any sense to have an effect on the enormous scale of the events that we have discovered in the ice core data.

Our Sun, as a plasma star

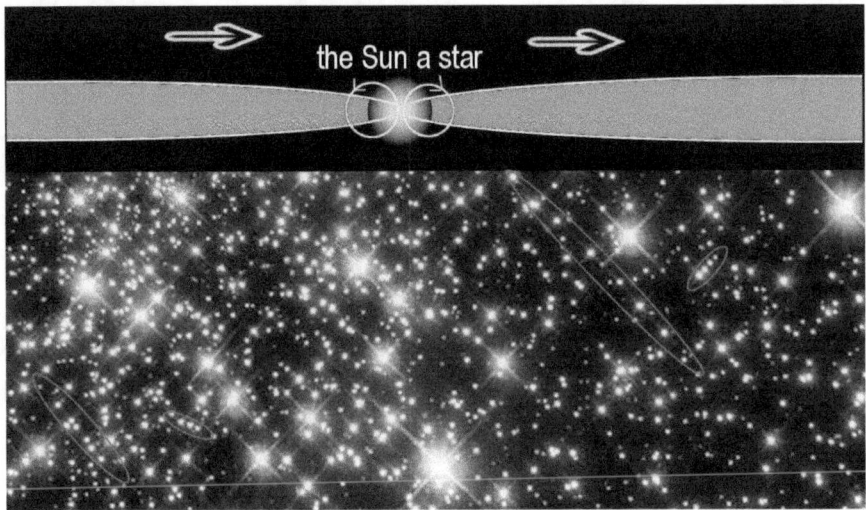

Our Sun, as a plasma star, is powered by interstellar plasma that is drawn together into plasma streams that typically tie the stars together into strings of various lengths, as the encircled examples show.

Plasma streams have a specific electric resonance

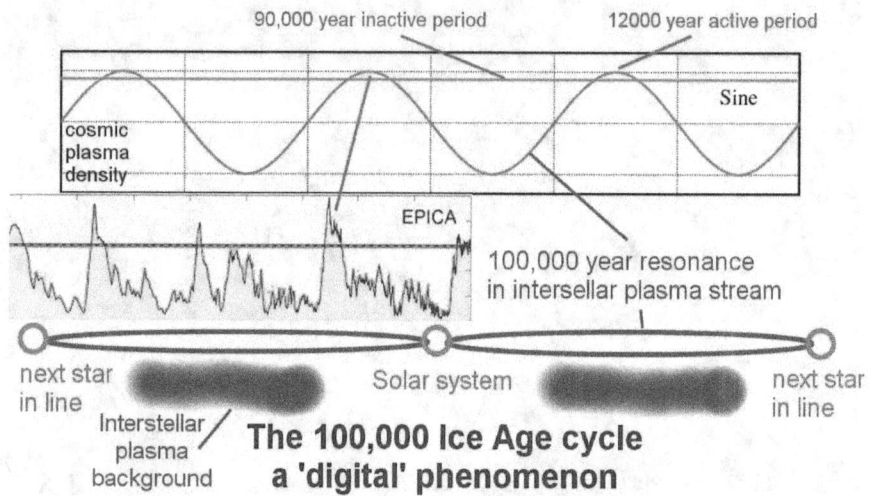

The 100,000 Ice Age cycle
a 'digital' phenomenon

The electric plasma streams also have a specific electric resonance,
according to the length of the plasma streams between stars. The
resonance determines the length of time of the ice age cycle. In the
solar system, an ice age cycle is presently slightly over 100,000
years in duration. It is made up of an interglacial warm period of
typically 12,000 years, in which the Sun is constantly powered, and
a 90,000-year glacial period in which the Sun is typically inactive.
In principle, the ratio between the active period and inactive period
is determined by the plasma-density threshold level, above which a
sun remains active, and below which it remains mostly inactive. The
active period becomes the interglacial period, and the largely
inactive period, the glacial period.

The orbital cycles of the Milankovitch theory

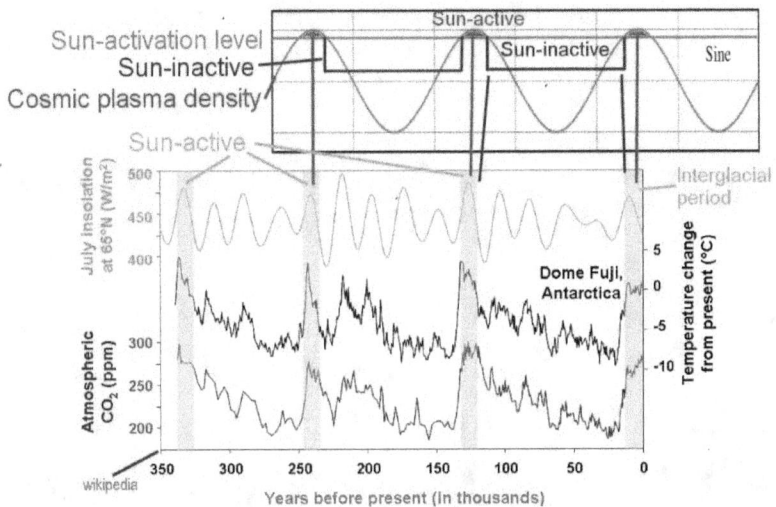

This means, that in principle, the timing of the ice ages is largely determined by the density baseline in the interstellar plasma streams, and by their resonance, or a combination of the in-flowing and out-flowing resonances interacting. In earlier time the ice age cycles were shorter than they are today, with a duration of 41,000 years. The shorter time suggests that a different resonance and a different baseline was dominant.

The orbital cycles of the Milankovitch theory, are evidently not causative factors for the ice ages, as their effects are too minuscule, but were themselves effects of the changing, immensely powerful, plasma-flow pattern.

The type of event that is big enough

The type of event that is big enough to cause the immense fluctuations, which nothing less could cause than the Sun going inactive, is totally possible for a Sun that is electrically powered, which it obviously is.

I have explored the factor of the electric Sun in a previous video. We have plenty of evidence on hand that the Sun is electrically powered, powered by plasma fusion occurring on its surface. We also have hard evidence on hand that a transition is in progress, towards the Sun going inactive in roughly 30 years.

And again this evidence takes us back to 1990 when the big ice core drilling was started. For the focus on the Sun itself, an exploration satellite had been launched in 1990 that had observed the Sun outside the 'fog' in the solar system's ecliptic.

In 1990 the Ulysses space mission was launched

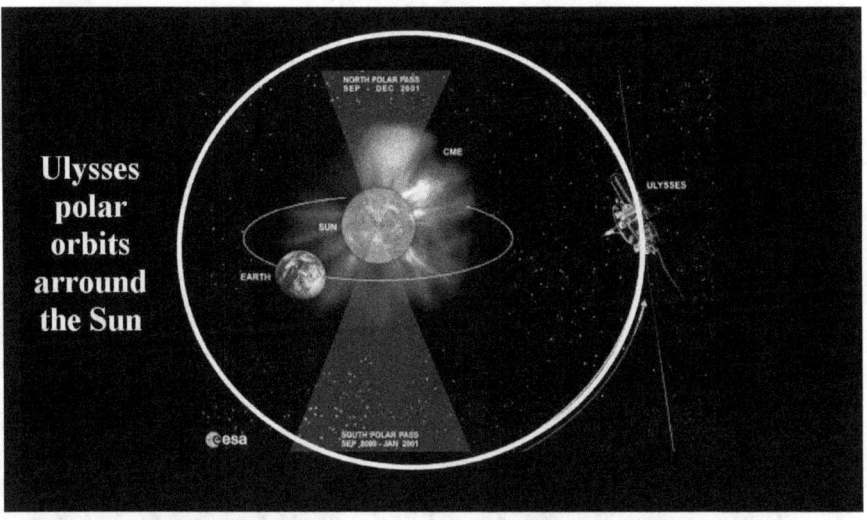

On the 6th of October in 1990 the Ulysses space mission was launched from NASA's space shuttle "Discovery." The Ulysses mission, placed a satellite into a polar orbit around the Sun that takes it far outside the ecliptic. The mission used the gravity of Jupiter to swing the satellite out of the ecliptic plane where solar measurements are inherently affected by the heliospheric current sheet that flows there. The deflection by the gravity of Jupiter has put the satellite it into an orbit perpendicular to the ecliptic where nothing disturbs the reading. The Ulysses satellite had reached Jupiter on the 8th of February in 1992. That's when its first solar orbit began.

A 30% reduction in the solar wind pressure

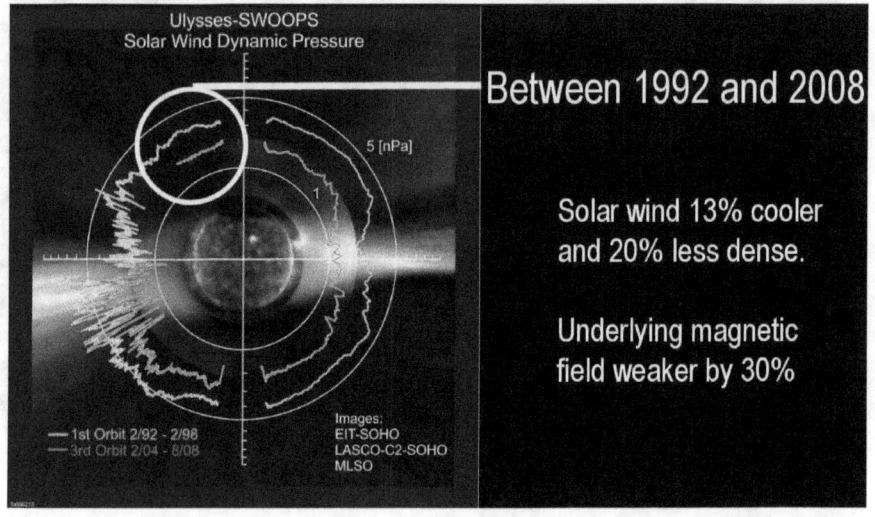

The satellite delivered data for the next 16 years until it was turned off during its third orbit. In these years of operation, it provided three major items of critical evidence. It measured over the span of its operation a 30% reduction in the solar wind pressure. That's a huge reduction for such a short period.

During its first path over the South Pole

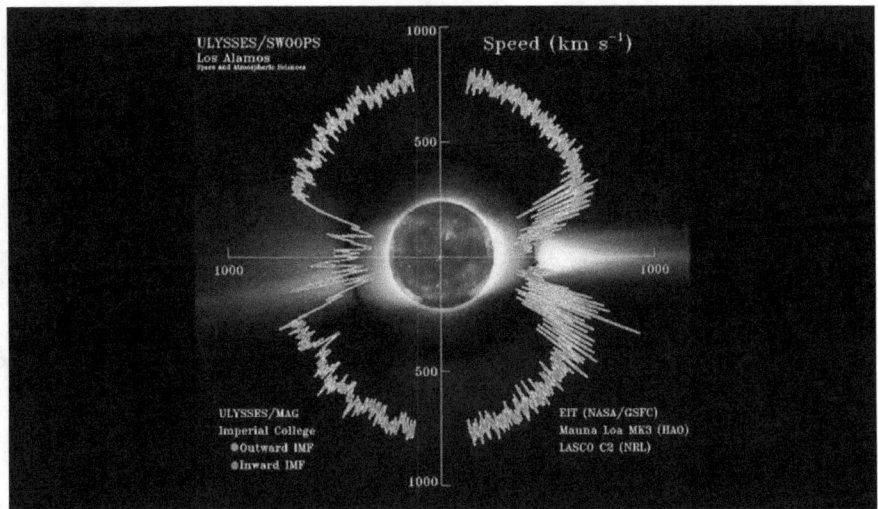

During its first path over the South Pole in 1994, the solar wind measurement was sharply disrupted. The interference left a gap in the measurements over the South Pole. The same happened again in 1995 over the North Pole of the Sun. The disruption phenomenon was observed for every path.

For an electrically powered sun

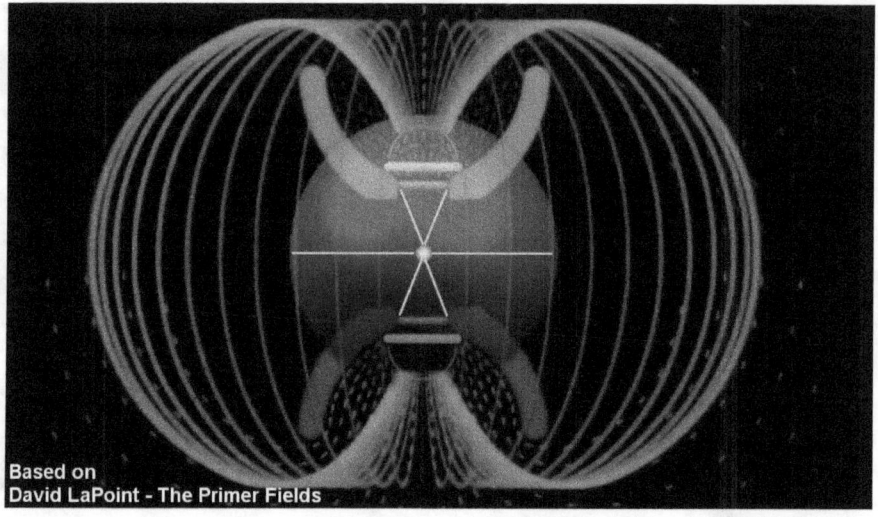

Based on
David LaPoint - The Primer Fields

The disruption may have come as a surprise, but for an electrically powered sun, the gap is exactly what one would expect to see. It thereby provides evidence that the Sun is electrically powered by plasma focused on it via two major primer fields.

In high-power laboratory experiments

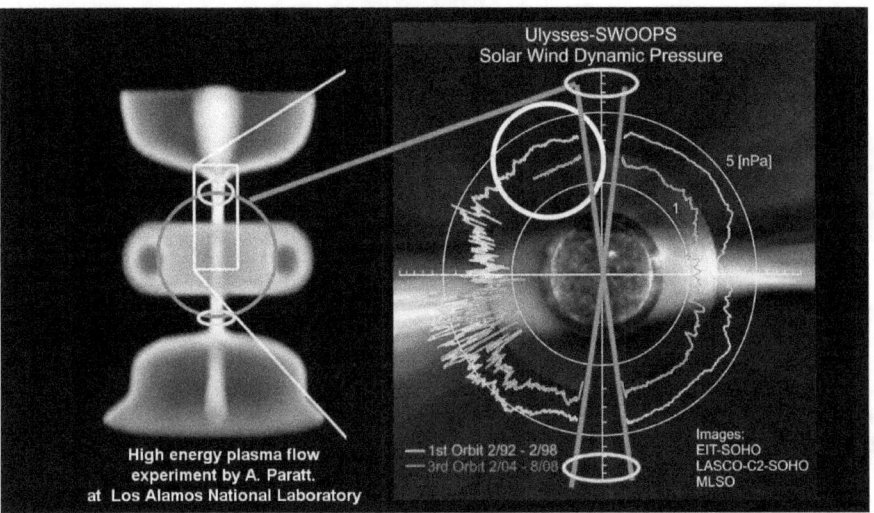

The disruption matches in principle the plasma-flow model that has been discovered in high-power laboratory experiments at the Los Alamos National Laboratory. The disruptions occur where the focused interstellar plasma streams would connect with the Sun, which they obviously do.

The astonishing 30% reduction in solar wind pressure

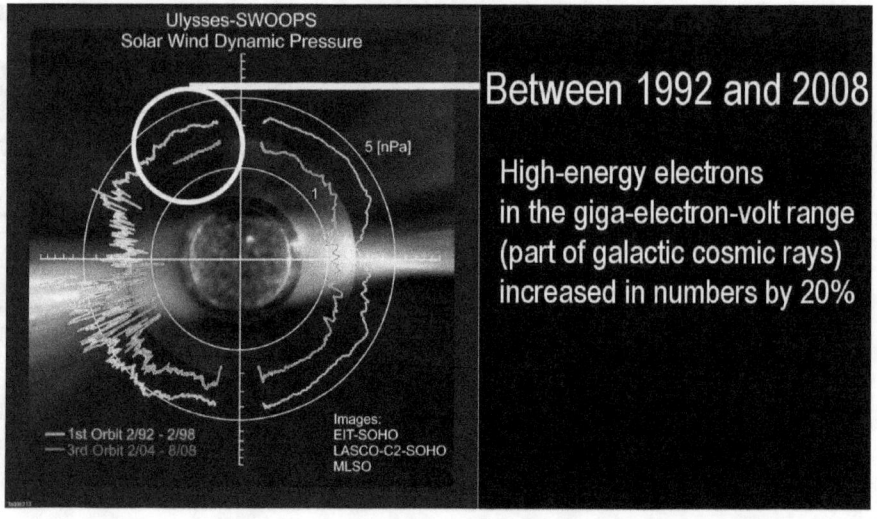

Ulysses-SWOOPS
Solar Wind Dynamic Pressure

5 [nPa]

— 1st Orbit 2/92 - 2/98
— 3rd Orbit 2/04 - 8/08

Images:
EIT-SOHO
LASCO-C2-SOHO
MLSO

Between 1992 and 2008

High-energy electrons
in the giga-electron-volt range
(part of galactic cosmic rays)
increased in numbers by 20%

The second major item of evidence that Ulysses has measured, is the astonishing 30% reduction in solar wind pressure between the first and third orbit. The surprise made headlines in the press. Shortly thereafter the mission was terminated. The satellite's transmitter was turned off, on the 30th of June in 2009.
The measured 30% drop in solar wind pressure in just 8 years, is enormous. In fact, the reduction is gigantic for such a short period.

Ulysses satellite had measured a 20% increase in Cosmic-Ray flux

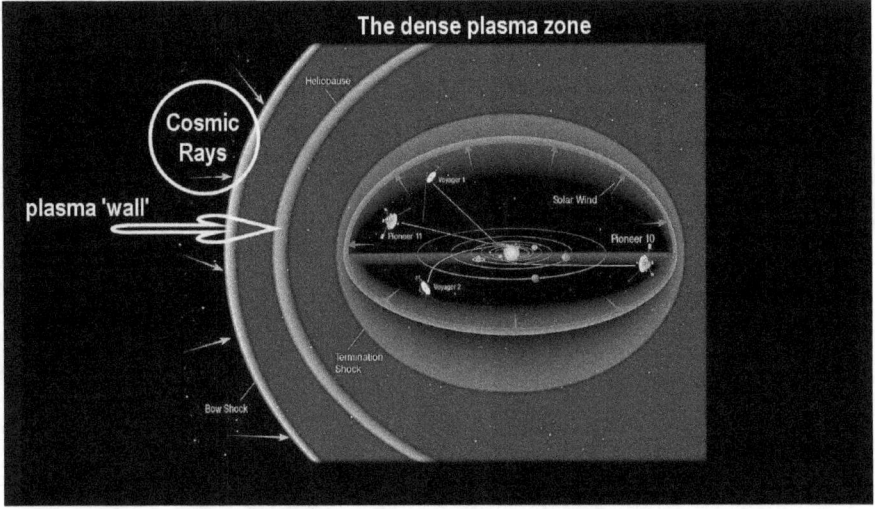

The reduced solar wind pressure had the secondary effect of weakening the shielding effect of the solar heliosphere around the solar system, which shields us against Galactic Cosmic-Ray flux. With the reduction in solar-wind pressure, the Ulysses satellite had measured a 20% increase in Cosmic-Ray flux over the same timeframe.

Solar wind measurements

Maximum temperature of liquid water at ambient pressure is 100 degrees Celsius: The Boiling Point

These are all hard items of evidence that big changes are in progress in the solar system. In an electric sun-system the solar wind can be likened to steam boiling off from a heated kettle. When the plasma input exceeds the density that the plasma-fusion reactions can consume, the excess pressure is vented off by the primer fields and becomes the solar wind. The solar wind measurements thereby become a measure of the 'health' of the solar system. When one sees a 30% drop in just eight years, such evidence becomes a cause for great concern.

If one projects the process forward

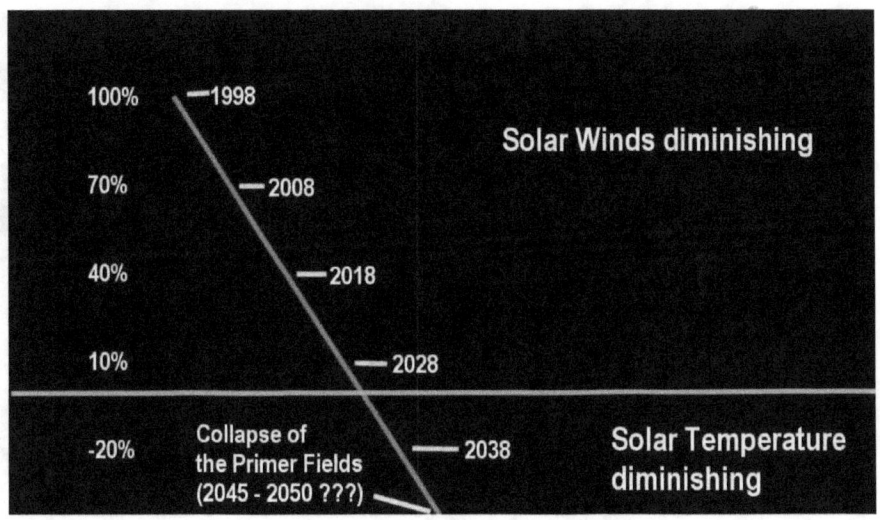

If one projects the process forward in a linear manner on the basis of a 30% drop in 10 years, the solar wind will likely have diminished to zero by 2030. From this point on the solar radiation itself would begin to diminish progressively towards the Sun going inactive at some point. The cut-off event will likely happen between the 2030s and the 2050s. Once this happens the magnetic primer fields collapse. It will take a big increase in plasma-flow density for the primer fields to form anew. During the last glaciation period the recovery of the Sun occurred in intervals of 1470 years.

Occurrences of the Red Sprite events

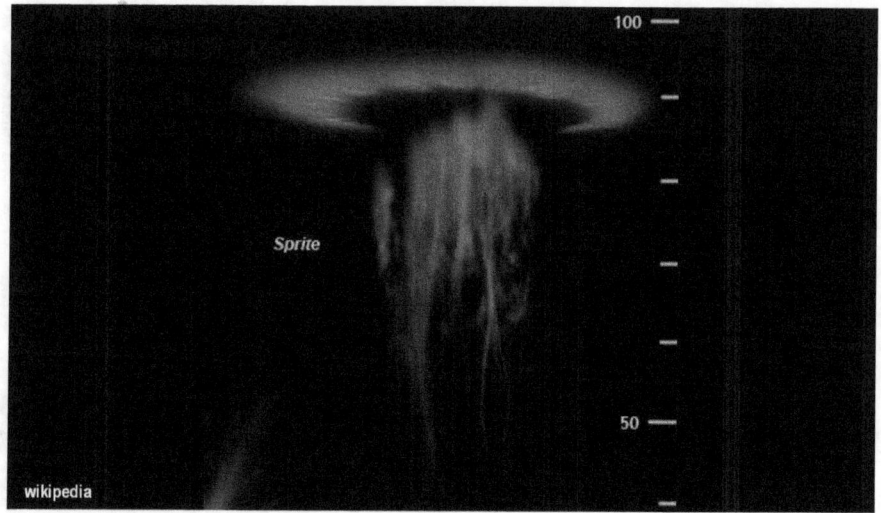

That primer fields can become established quickly when the conditions are right and vanish quickly when the conditions are exhausted, is evident in the occurrences of the Red Sprite events that occasionally become visible in the tropics high above large storms clouds. The sprite events fair up, but rarely last for more than a second. The larger event of our interglacial Sun has so far lasted for roughly 12,000 years, which is typical for interglacial periods.

The Sun goes inactive like a sprite turning off

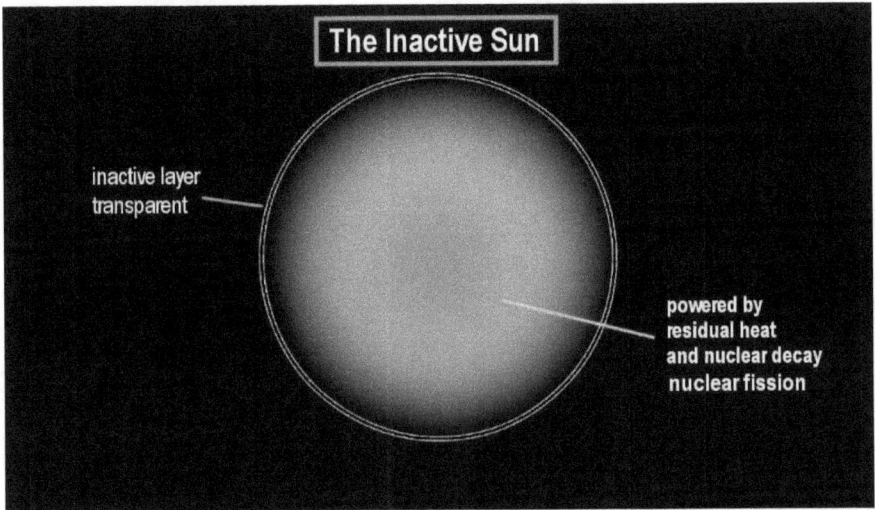

Once the Sun goes inactive like a sprite turning off, it will likely continue to glow for a long time by residual processes, although at a much-reduced energy level, at probably 30% of the present level. The resulting enormous reduction in radiated energy would certainly be large enough to cause the gigantic Ice Age cooling that the ice cores tell us of.

The Dansgaard Oeschger oscillations

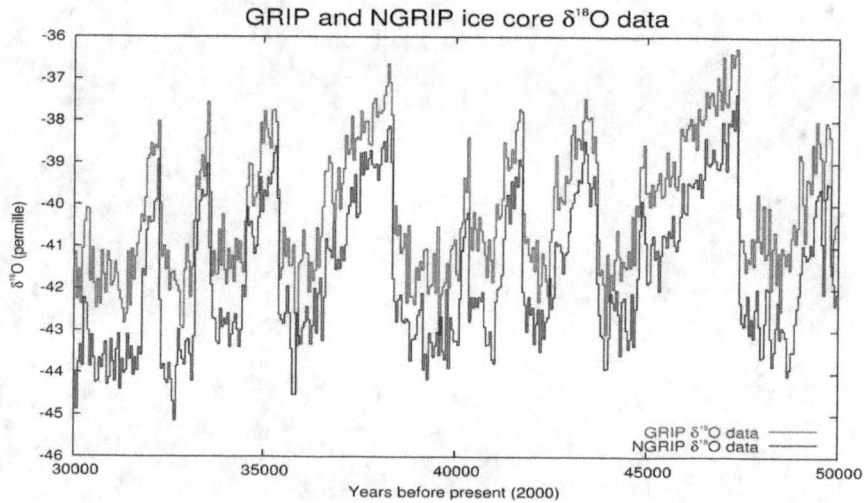

GRIP and NGRIP ice core δ¹⁸O data

The shock effect of the solar-fusion consumption of plasma being interrupted, may eventually recover the incoming plasma density enough for the Sun to start up again for a short period. The periodic nature of the Dansgaard Oeschger oscillations, suggest that such off/on/off events were happening throughout the last Ice Age, with the results as shown here.

This is the kind of rapid transition in the earth environment that we need to prepare our world for, which we need to complete in the few years of interglacial time that we may have remaining, possibly 30 years, if that.

The lead time is short

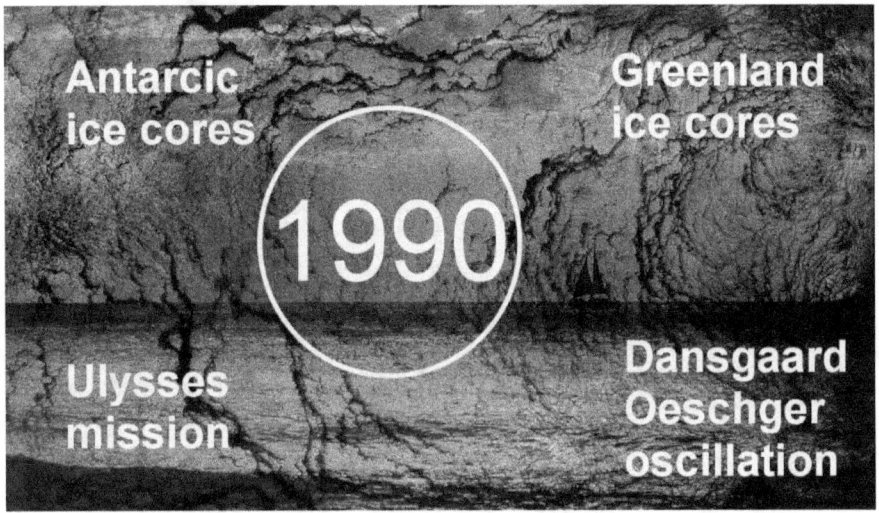

The lead time that we may still have is short. It is short, because the entire discovery process of the ice age dynamics didn't begin until the early 1990s when the deep ice core drilling began and the Ulysses satellite was launched to explore the connection with the Sun.

Nobody had connected the two events

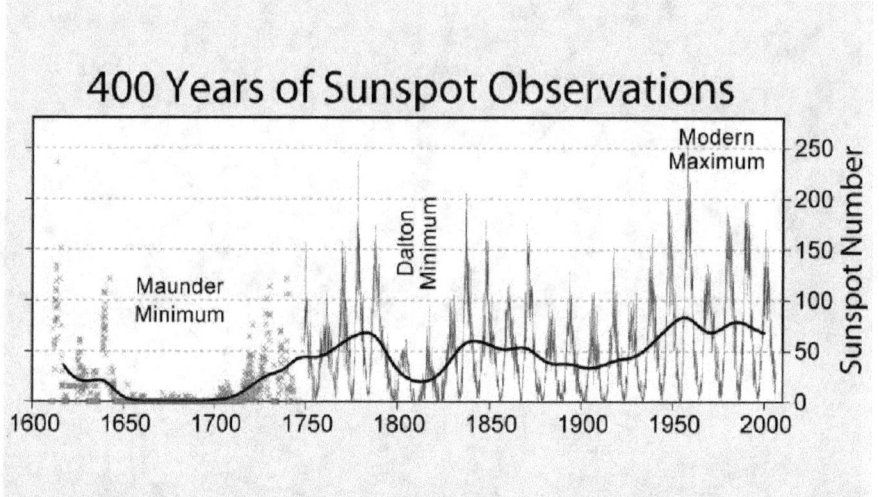

The 'earlier' connection between the weaker Sun of the Maunder Minimum, and the colder climates at the time, had been on record, and fully out in the open, back in the 1600s, but no one had seen the connection. Nobody had connected the two events then.

Will we connect the dots?

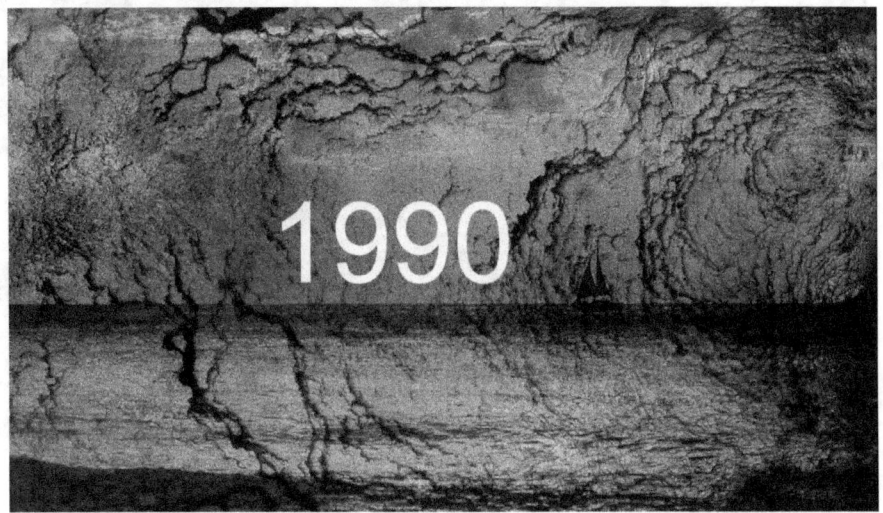

We are back in the same boat again from the 1990s onward. The evidence of big events stands before us. The ice core evidence is plain. It has been measured in details and the measurements were recorded. The Ulysses spacecraft saw the solar system getting dramatically weaker while it orbited. But will we connect the dots?

A third set of evidence to jolt us

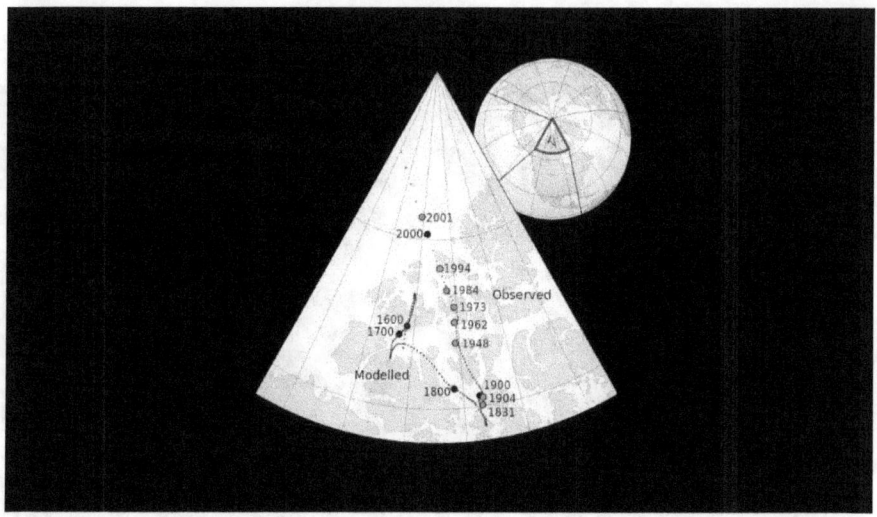

It is as if we needed a third set of evidence to jolt us to attention. This third set of measurements was produced on the ground in the high arctic of Canada in the form of magnetic measurements that tracked the Earth's magnetic pole 'drifting' northward. This is significant. But what does it mean? It means that we see evidence here of a dramatic weakening of the magnetic strength of the primer fields that are affecting the Earth's magnetic dynamo effect.

The magnetic effect of the spinning of the Earth

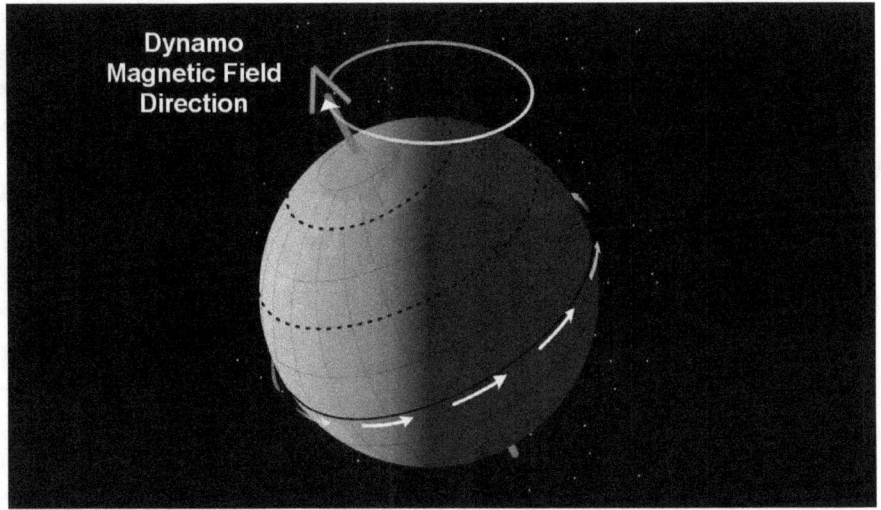

The magnetic effect of the spinning of the Earth, by the principle of magnetic fields, causes the magnetic field lines to be formed along the spin axis of the Earth. If no other factors are applied, the Earth magnetic North Pole becomes identical with the Earth geographic North Pole.

Within the giant sphere of the primer fields

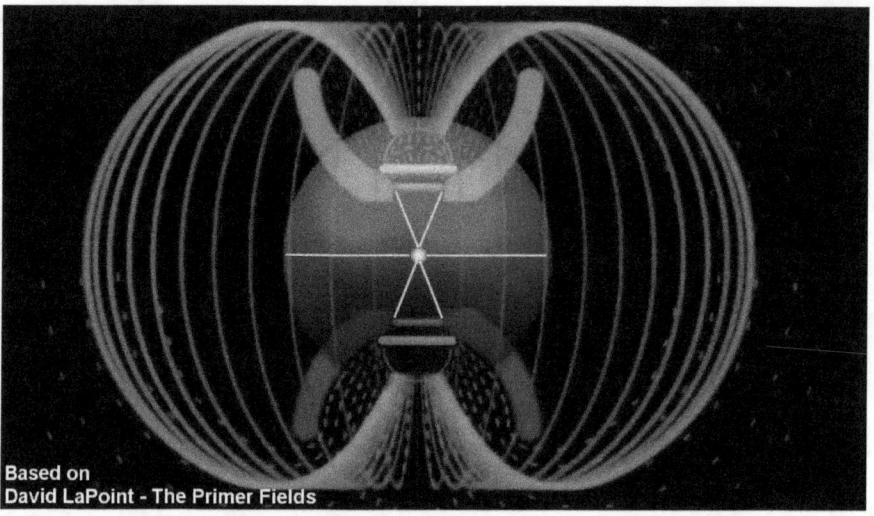

However, if one considers that the Earth orbits within the giant sphere of the primer fields, a second magnetic force affects the Earth.

The second force has a different orientation

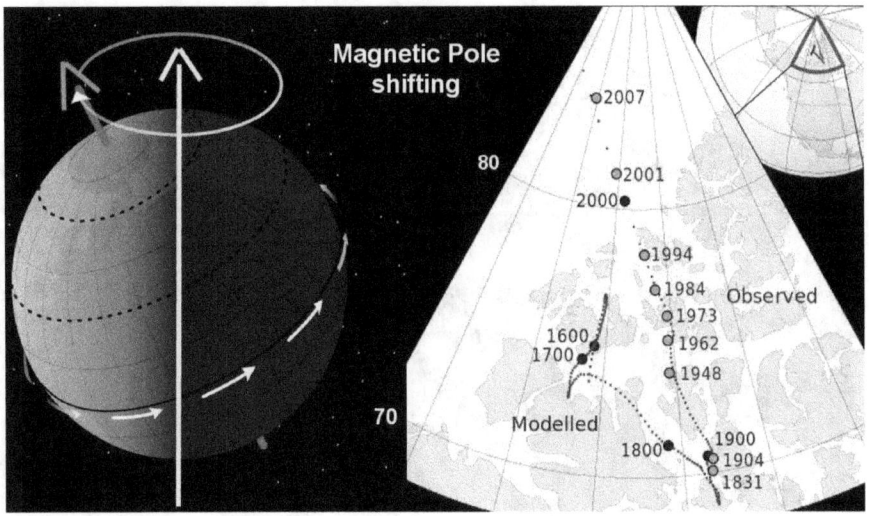

If the second force, which has a different orientation, is extremely dominant, it will pull the effective magnetic pole into its direction, for a maximum deflection of 23 degrees according to the tilt of the spin-axis of the Earth. In 1831, at the time when the Earth recovered from the Little Ice Age, the measured deflection was 20 degrees. This was a large deflection, caused by strong primer fields. From this time on, slowly at first, the deflection of the Magnetic North Pole, as it was measured on the ground, drifted closer and closer towards the geographic pole. The measured drift tells us that the primer fields were weakening, whereby the dynamo field became increasingly more dominant.

A large weakening of the solar system

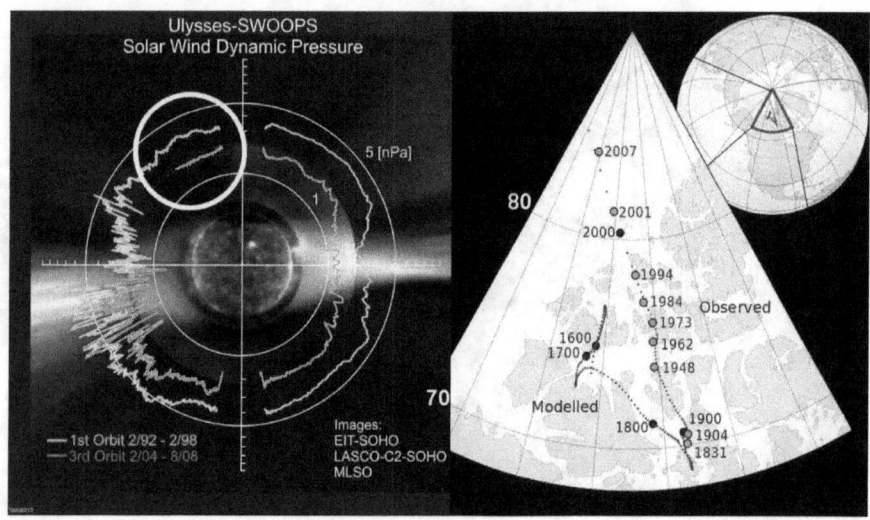

That the magnetic pole drift was not caused by the Earth itself is evident by the corresponding weakening of the Sun that had been measured by the Ulysses spacecraft between 1992 and 2008. We are evidently looking at a large weakening of the solar system being in progress.

The weakening of the solar system began earlier

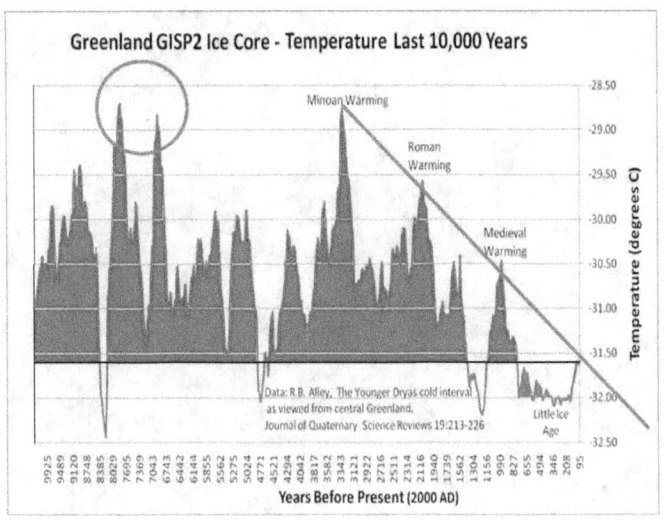

Of course, as the Little Ice Age tells us, the weakening of the solar system began already much earlier in time. The ice core samples from Greenland tell us that the weakening of the solar system began as far back as 3000 years ago, slowly at first, but has gradually accelerated.

The dramatic, fast weakening that Ulysses saw, which is evident also in the magnetic pole drift, suggests that a terminal phase has begun.

Hasn't the solar system recovered from the Little Ice Age?

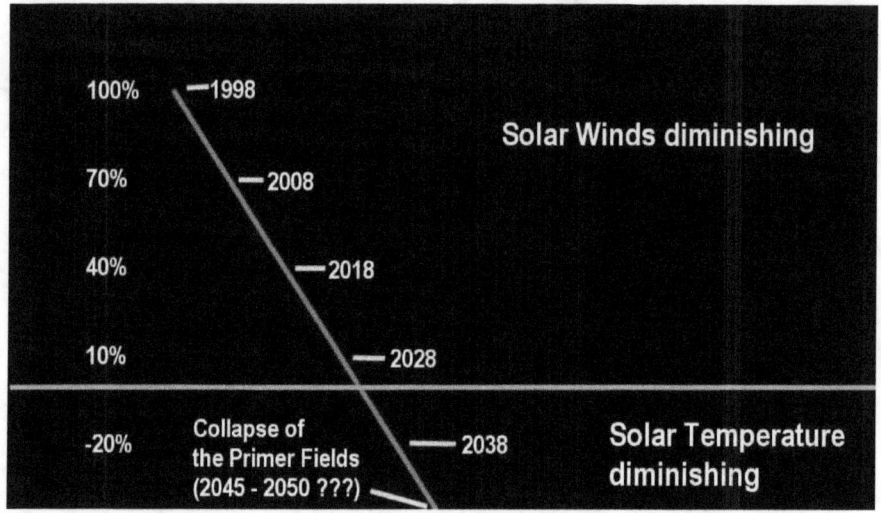

It is tempting to hope that the presently indicated trend will somehow reverse before the Sun goes inactive. Hasn't the solar system recovered from the Little Ice Age? Yes, it has recovered, but it recovered from a stronger base.

Stronger deflection of the magnetic pole

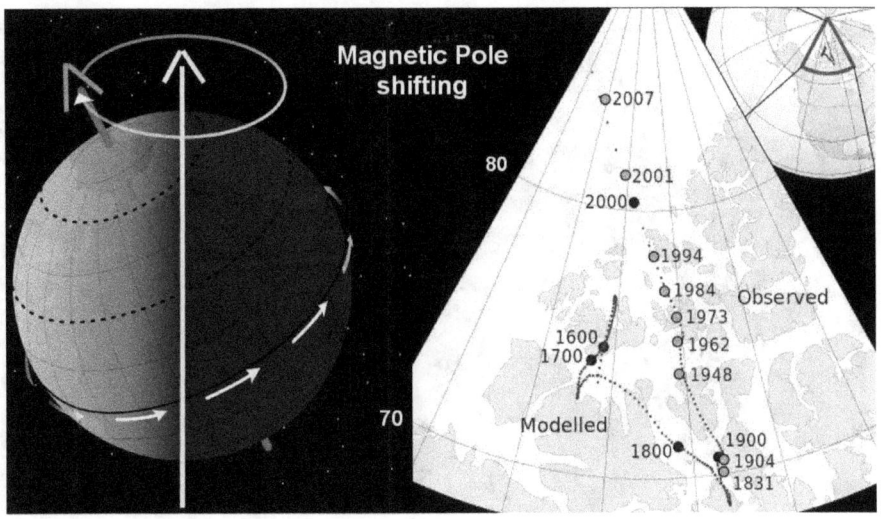

The stronger base is evident in the stronger deflection of the magnetic pole by the primer fields. The deflection at the worst of the Little Ice Age hadn't drifted nearly as close to the geographic pole than we have it today. This means that the primer fields were far stronger then. This apparent paradox tells us that the effective field strength of the primer fields has not a singular source, but is the effective composite of three nested primer fields overlaid upon another.

Theorized in 1932 by Jan Oort

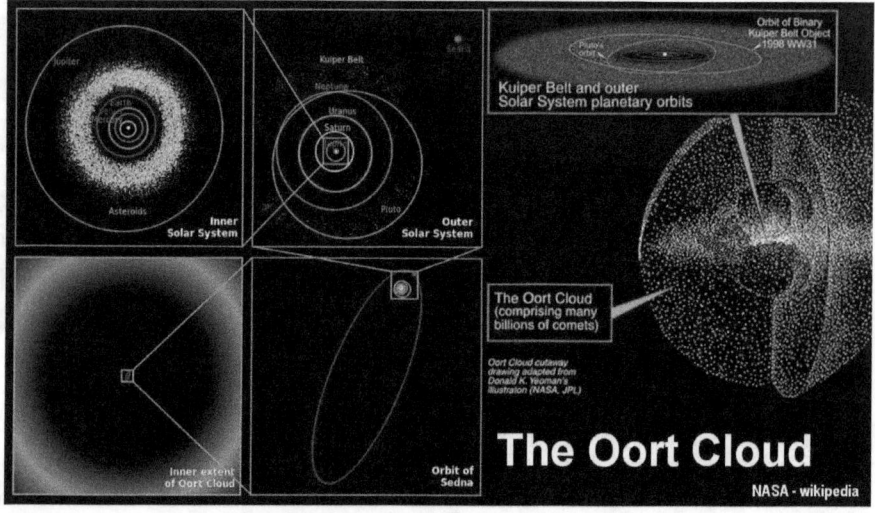

It was theorized in 1932 by the Dutch astronomer Jan Oort and the Estonian astronomer Ernst Öpik, that a spherical cloud of asteroid objects surrounds the solar system up to a distance of close to a light year.

The cloud is theorized to be made up of two nested parts, an inner part that extends for less than a third of it, and a larger outer sphere. The Sun, located deep within it, is just a tiny speck.

A three-fold nested complex

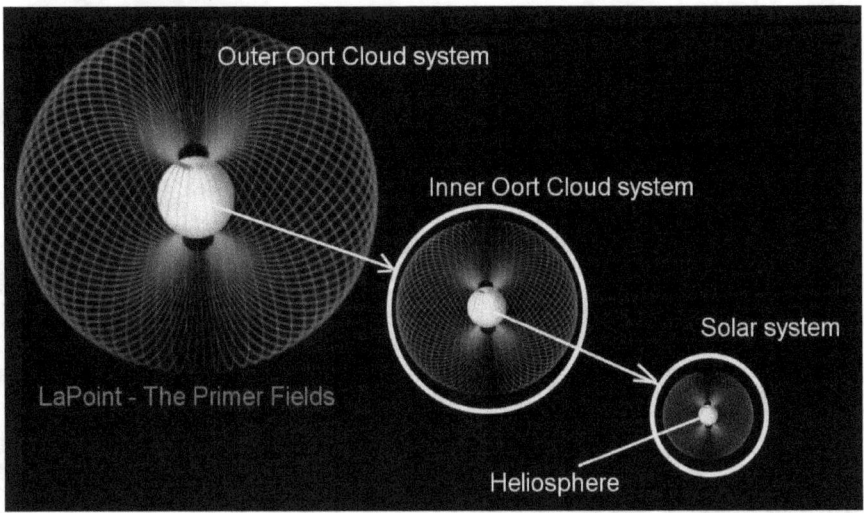

If one regards this two-part giant phenomenon as a part of a large plasma flow system, and plasma concentration system, then we can recognize the existence, in principle, of a three-fold nested complex of primer fields with each having a correspondingly longer resonance characteristic.

The outer Oort cloud, by its size

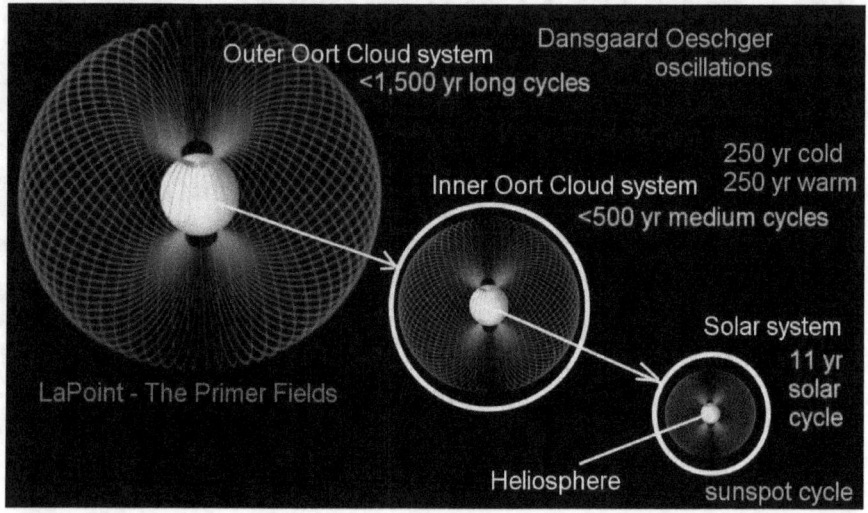

The outer Oort cloud, by its size, would likely have a resonance that corresponds with the Dansgaard Oeschger oscillations during the glaciation period.

The long resonance of the outer Oort cloud

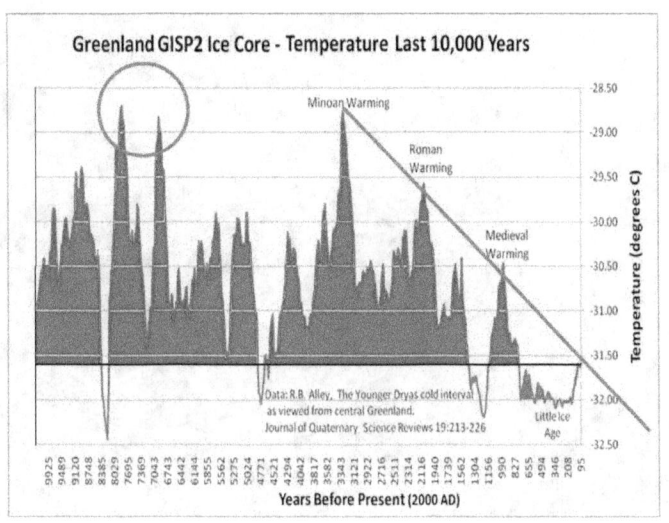

The long resonance of the outer Oort cloud may also be a causative element for the big spikes that we see as warming events in interglacial time.

The resonance of the inner Oort cloud

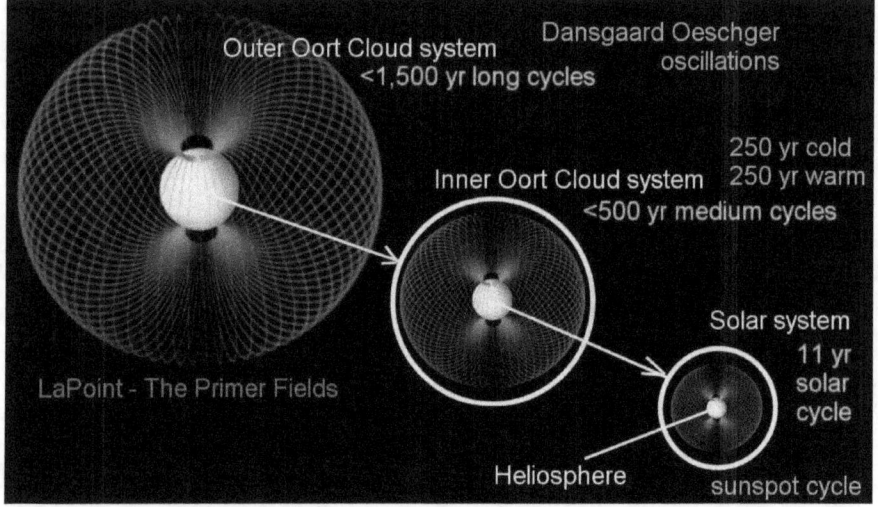

The resonance of the inner Oort cloud, being much smaller in size, would be reflected in the shorter cyclical events such as we see in the 250-year cooling of the Little Ice Age, and the subsequent 250 years re-warming afterwards. The still-strong magnetic pole deflection during the Little Ice Age, suggests that there was still a lot of strength left in the outer Oort cloud, which we don't have this time around.

Before the current 250-year down-cycle has ended

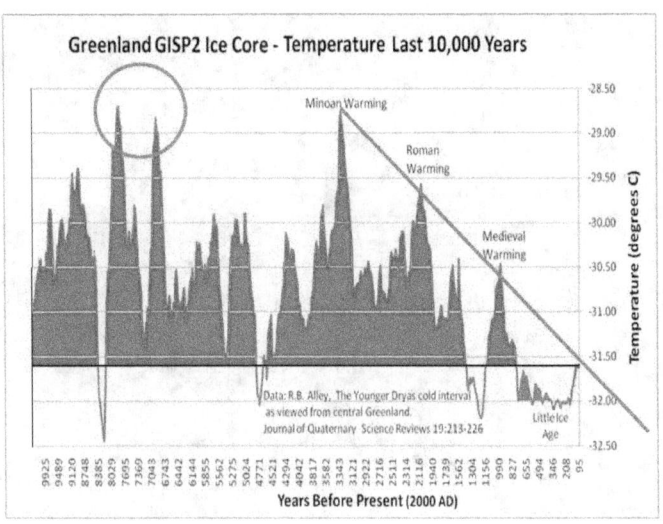

This is what the green line indicates. It means that the 250-year down-swing that has now begun, is beginning in a dramatically weaker environment. Without the strong backup that still existed during the Little Ice Age, the Sun will likely go inactive this time around, long before the current 250-year down-cycle has ended.

Society is foolishly gambling with its very existence

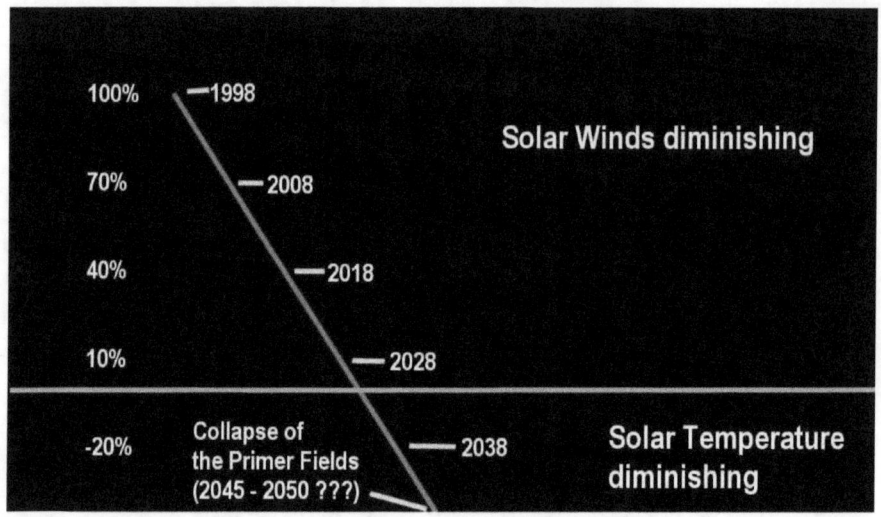

It is more likely that the diminishment of the solar wind pressure will accelerate instead of being reversed by a strong backup. No evidence exists that a backup is in the works for a significant reversal to take place.

Society is foolishly gambling with its very existence, by assuming that a reversal of the current dynamics will magically happen that would justify its refusal to prepare the world for the impending Ice Age transition in roughly 30 years.

This is an insane gamble

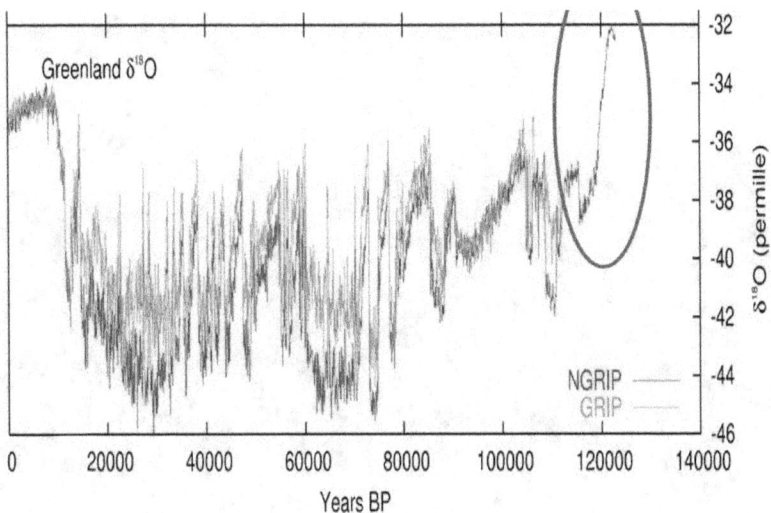

Greenland δ¹⁸O
NGRIP
GRIP
Years BP

The current gamble that that the near Ice Age will pass us by, is a gamble that all evidence tells us, we will loose. This is an insane gamble. To put it bluntly, the evidence tells us that we are on the fast track to an icy hell. The last ice age transition, measured in the ice of North Greenland, gives us a hint of what we are up against. When all of this evidence is fitted together in the mind into a singe package of science perception, which takes us beyond what the eye can see, alarm bells should ring. The alarm bells should signal that the final phase towards the next glaciation event is most likely in progress, towards an inactive Sun, and that no form of gambling with the future existence of humanity is justified.

But why aren't the alarm bells ringing?

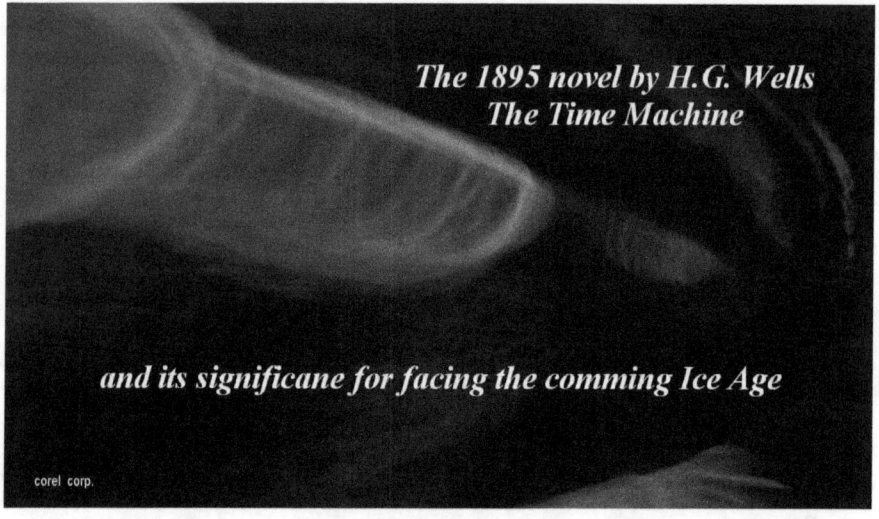

The 1895 novel by H.G. Wells
The Time Machine

and its significane for facing the comming Ice Age

corel corp.

But why aren't the alarm bells ringing? The simple answer is that science is being depressed for political objectives by the masters of the world empires who have waged war against scientific progress for centuries. The reason was stated bluntly, and repeatedly, by the masters of the ruling empire. It has been stated that scientific progress is the greatest danger to any feudal oligarchic system of empire, because empire exists by the looting of society. Empire never stands on its own merits. It has none. Empire protects itself by keeping society small, impotent, and its science depressed. The details involved are too many to list here, however a few points of evidence in modern time are noteworthy.

The scientific community became concerned

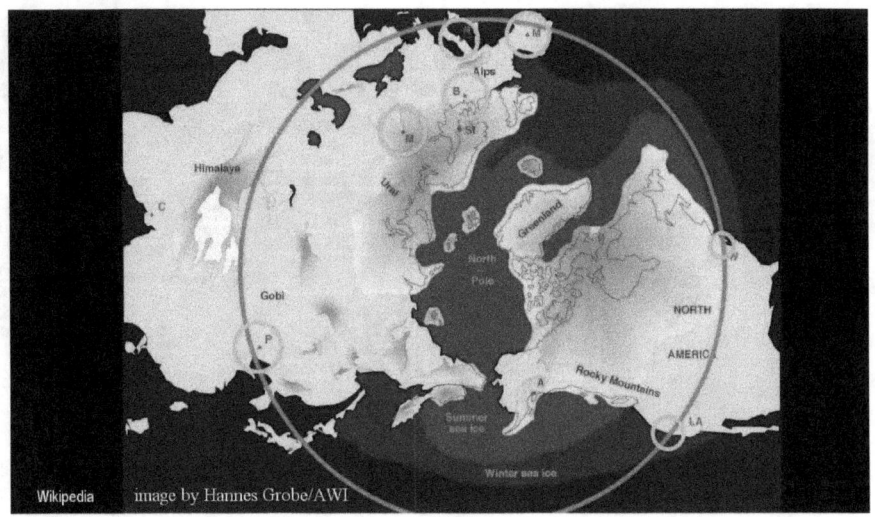

One point to note is, that the scientific community became concerned already in the early 1970s about what must be done to protect humanity against the approaching Ice Age. The scientific community called for a world forum to discuss the subject.

The world forum was convened in 1974

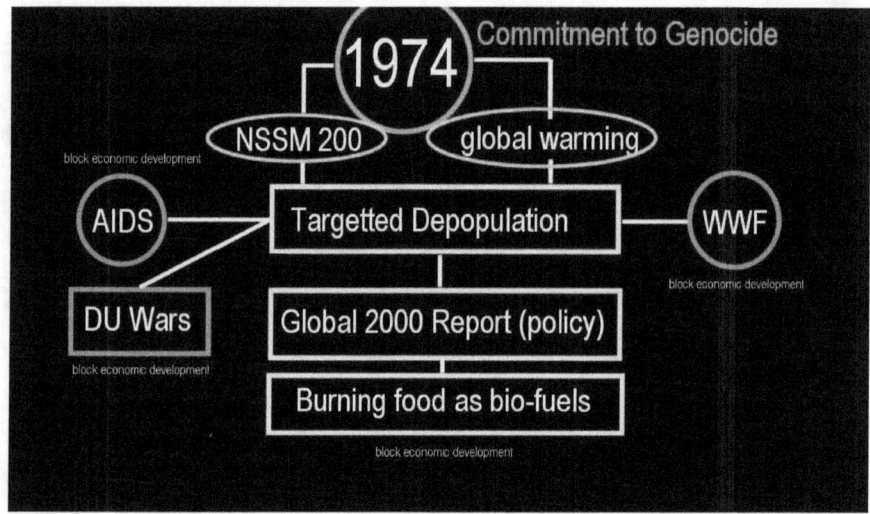

The world forum was convened in 1974, however, it was staged in upside down form as the U.N. Conference on Population in Bucharest where the announcement of the global warming doctrine was made, for which no evidence really exists, but which was immensely promoted thereafter.

In the shadow of the global warming doctrine that was essentially a hoax, the super-secret infamous NSSM-200 policy was quietly concocted in the same year, 1974, which targeted third world nations for controlled depopulation. The timing coincidence strongly colors the real intention for the global warming doctrine and takes it far outside the realm of science. Today, 30 years later, the global warming doctrine is murdering up to 100 million people a year with starvation by the mass-burning of food in the form of biofuels.

The scientific community protested

we want to be free

Scientists Appealing the Global Warming Doctrine

The 1992 Heidelberg Appeal
The 1997 Leipzig Declaration
The Oregon Petition Project
Open letters to the Prime Minister of Canada
The 2007 U.S. Senate Report
100 prominent international scientists warn the U.N.
The new (2008) Oregon Petition Project (ongoing)

corel corp

The scientific community protested against the perpetrated hoax that was barely becoming recognized then for what it was. The science community responded with numerous petition projects to inspire the leaders of the world to act on truthful scientific evidence instead of on contrived imperatives.

The largest of the resulting opposition projects garnered 17,000 signatures from the scientific community worldwide, but those became promptly ignored in the political world as the truth really wasn't on the agenda.

One opposition project was launched by the US Senate

The U.S. Senate Report

400 detailed individual declarations online

For details see: ice-age-ahead-iaa.ca

One opposition project that you have probably never heard of either, was launched by the US Senate. The project published 400 detailed statements of scientific opposition to the global warming doctrine, presented by concerned individuals with an academic degree. This massed voice too, was simply ignored.

In Lies We Trust

A dozen years of lies

Lord Christopher Monckton exposes the climate fraud on the world stage.

thousands of leaked e-mails w. evidence of fraud, falsified 'official' data destruction of historic data

computer division at the Climate Research reveals programs and data in a hopeless, tangled mess so that global temperature trends were simply made up.

COP15 COPENHAGEN
UN CLIMATE CHANGE CONFERENCE 2009

Global warming proponent Stephen Schneider; ". ..[W]e have to offer up scary scenarios, make simplified, dramatic statements, and make little mention of any doubts we may have. Each of us has to decide what is the right balance between being effective and being honest."

Nov. 1997 - 21st Century Science and Technolgy

Lies exposed at the U.N World Conference

One promoter of the political hoax project stated that each one in the scientific community must decide for himself were the balance lies between being 'truthful' and being 'effective.' On this platform, a banner has been erected high over the land of science, that reads, "In Lies We Trust."

Sanitized by omissions or 'corrections'

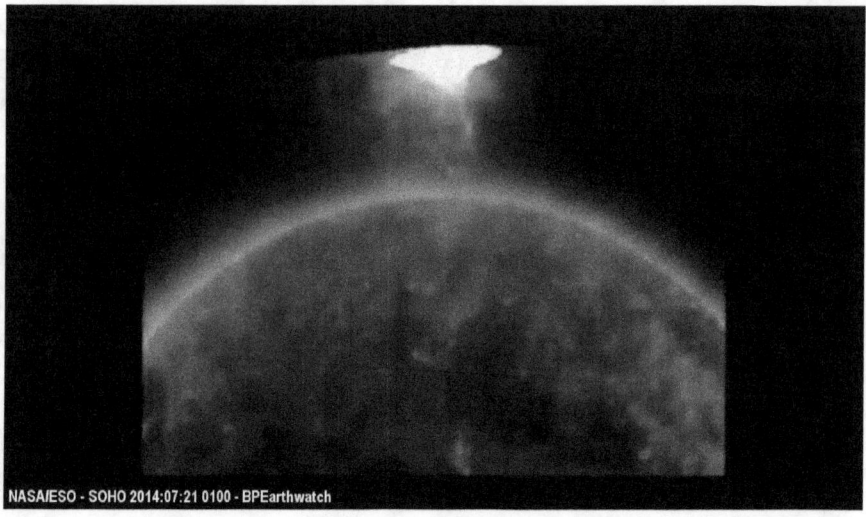

NASA/ESO - SOHO 2014:07:21 0100 - BPEarthwatch

Under the established banner many a published scientific data has evidently been sanitized by omissions or 'corrections'. Sometimes, however, 'the pollution of the truth' makes it briefly past the guards, and is seen by the public, as did the image shown here of a gigantic event over the North Pole of the Sun photographed by the SOHO spacecraft.

To assure 'the purity of the pabulum'

NASA/ESO - SOHO 2014:07:21 0100 - BPEarthwatch

This image, with all its different spectra, was deleted from the published data on the next day. It is unknown to what extend published scientific data is routinely sanitized in order to assure 'the purity of the pabulum' that is dished up for society.

Often the pabulum is sweetened with fairy-tale dust

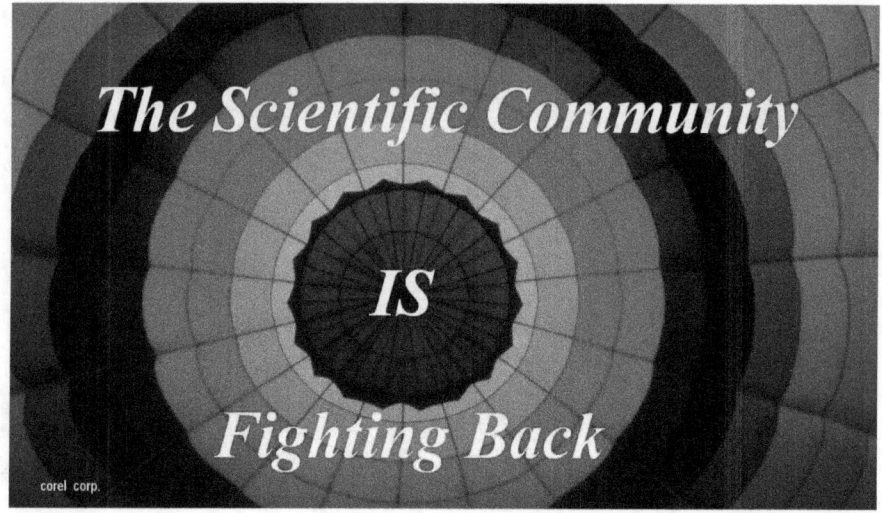

Often the pabulum is sweetened with fairy-tale dust. It was revealed, for example, during the time of the 2009 Climate Change Conference, that, based on the publication of several hundreds of emails collected from the servers of relevant institutions, that the bulk of the data that is fed to the public in support of the global warming scare, has simply been made up.

One of the early activists

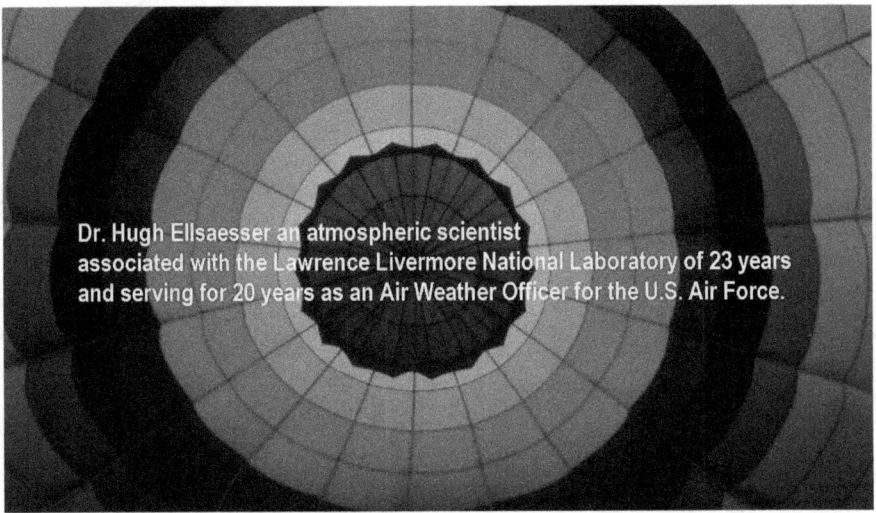

Dr. Hugh Ellsaesser an atmospheric scientist associated with the Lawrence Livermore National Laboratory of 23 years and serving for 20 years as an Air Weather Officer for the U.S. Air Force.

One of the early activists on the front for truth in science, was an atmospheric scientist who had been associated with the Lawrence Livermore National Laboratory for 23 years and had served for 20 years as an Air Weather Officer for the U.S. Air Force. He describes in an article how the big international protest movements against false science had developed from 1992 on. This occurred with the big ice core drilling going on in the background.

The great protest movements

we want to be free

**Scientists Appealing
the Global Warming Doctrine**

The 1992 Heidelberg Appeal
The 1997 Leipzig Declaration
The Oregon Petition Project
Open letters to the Prime Minister of Canada
The 2007 U.S. Senate Report
100 prominent international scientists warn the U.N.
The new (2008) Oregon Petition Project (ongoing)

corel corp.

The first big opposition movement in the science community, which most people never heard of, started in Germany in 1992 at the University of Heidelberg. It was followed up 5 years later at the University in Leipzig, and then again with a major effort organized in Oregon in the USA, and so on.

The great protest movements had produced over 50,000 signatures and statements from the scientific community from numerous countries, with numerous Nobel Laureates among the respondents.

The Heidelberg Appeal, that went out to the scientific community in 1992, was responded by over 4,000 scientists, including 69 Nobel Laureates from 69 countries. It was hoped that such a wide-based voice would have an impact on the policy makers at the Rio Earth Summit. Of course, since the truth was not on the agenda there, the voices from the science community weren't even heard.

Dr. Ellsaesser states that over 35 organizations had been publicly standing up against the global warming doctrine at the time, whose voices were simply ignored.

The Leipzig Declaration

The Leipzig Declaration had solicited only meteorologists and climate specialists for a completely focused statement on the issue. It had collected 110 signatures from this highly specialized group. The resulting statements by leading-edge specialists in the field, was subsequently submitted to the Kyoto climate conference, but, there, as it had already become the norm by then, their voices were simply ignored.

A global suicide pact

In the shadow of this ban of the truth, the Oregon Petition Project against the global warming doctrine was launched. It was launched as a protest against the Kyoto agreements that were referred to by Russia's Academy of Science, as 'a global suicide pact.'
The Oregon Petition project went out to all scientists worldwide, to gain their written support for a campaign to urge the nations of the world to reject the Kyoto agreements. The world responded with 17,000 signatures from scientists, mostly with high degrees, standing in opposition to the unscientific assumptions of the global warming dogma. The project seemed to have succeeded significantly, as only a few nations have actually ratified the Kyoto agreements.

The big 2007 Senate Report

As I already noted, other protests followed in later years. One of these was the big 2007 Senate Report, which presents a collection of over 400 detailed opposing submissions, mostly from the academic community, each one outlying the individual's reason for opposing the global warming dogma. Their statements are all online. But by then, the issue of global warming, as a world issue, was already essentially dead, except as a political issue of insanity

The 2009 Copenhagen Climate Change conference

The 2009 Copenhagen Climate Change conference, that was designed to revive the global warming hoax, promptly failed spectacularly. It collapsed without a single agreement. It had drifted insanely far into the realm of purely imperial objectives with no scientific foundation for anything, to support anything. At the end India simply walked out in disgust.

Hopefully, the still remaining forced opposition against the ice age dynamics, will some day soon, similarly, fizz out into nothing, as we are running out of time on this front.

Professor Zbigniew Jaworowski

Zbigniew Jaworowski, PhD, MD, D.Sc.
chairman of the Scientific Council
of the Central Laboratory
for Radiological Protection in Warsaw

"CO2 - the greatest scientific scandal of our time"
for details, see: ice-age-ahead-iaa.ca

... a world-renowned atmospheric scientist
and mountaineer, who has excavated ice
out of 17 glaciers on 6 continents in his
50-year career.

21st Century Science and Technology

Among the voices in the scientific community, warning of the return of the Ice Age, was the world-renowned atmospheric scientist and mountaineer, Professor Zbigniew Jaworowski, who also served as chairman of the Scientific Council of the Central Laboratory for Radiological Protection in Warsaw'. He wrote in a paper in 2003, that the transition period to the deep cold temperatures can be as short as 1 to 2 years, and can begin virtually without warning, or that the transition might unfold slowly over a span of 50 years. The professor speaks to us as a world-renowned atmospheric scientist who has excavated ice out of 17 glaciers on 6 continents in a 50-year career. He warns us that the cyclical Ice Age transition is already half a millennium 'overdue.'

Get the fur coats out of the closet

For the timing, Professor Zbigniew Jaworowski, quotes other sources in his paper, that suggest that the transition may occur between 50 to 150 years from the present. More accurate forecasts were hard to make at the time, since the dynamics of primer fields had not been discovered in 2003. However, the professor points out at the end of his paper, that continuous on-the-ground temperature measurements taken at the Solar Terrestrial Institute in Irkutsk in Southern Siberia, have registered a sharp drop in annual average temperature, beginning in 1998, which became later confirmed by the solar measurements of NASA's Ulysses mission, and by corresponding measurements of magnetic pole drift in the high arctic of Canada and beyond.

And so, Professor Jaworowski. simply suggested, 'The Ice Age Is Coming', 'Get the fur coats out of the closet.'

The coming Ice Age is definitely not off the horizon

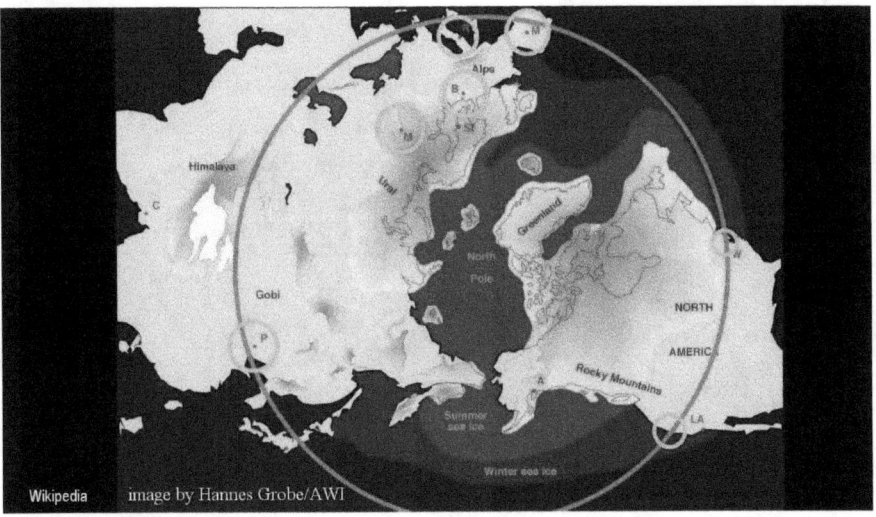

The coming Ice Age is definitely not off the horizon in scientific thinking. It is just beginning to unfold. The subject is merely pushed off the stage for the time being, by political forces in defence of the western oligarchic imperial system that has no natural foundation to exist, and thus cannot exist in the light of honest scientific development, especially the huge development that is mandated by the ice age dynamics.

The enormous economic development that is required

The enormous economic development that is required to prepare the world for human living under an inactive Sun, with 6000 new cities being built in the tropics for a million people each, mostly afloat, all with completely new industries and agriculture alongside, and with everything being constructed in the short span of 30 years, would sweep the imperial looting system out of existence. For this reason, the very notion of the Ice Age Challenge is being resisted, and the related fields of evidence are being 'sanitized' into oblivion.

Why have no such events been detected

The result that we see here may answer the question that is sometimes asked about the near potential of the Sun going inactive and the lack of evidence for it in the sky. People ask, "why have no such events been detected with telescopes, or by the Kepler spacecraft that searches for faint intensity variations in stars as evidence for planets orbiting the distant stars? The mission searches for the faintest changes in light density. Wouldn't such a mission see a sun going inactive, somewhere?

The reason why no such events have been reported, may simply mean that these cases have either been sanitized out of existence by the planet-scanning computer software that is not programmed for large changes, or that such occurrences are sanitized by the parameters that limit the reporting of what is seen.

It may also be that not a single sun, of the 145,000 stars that the Kepler mission had monitored in its narrow field of vision, has gone inactive during its slightly over two-years of operation.

Solar cut-off events may not happen for a few decades

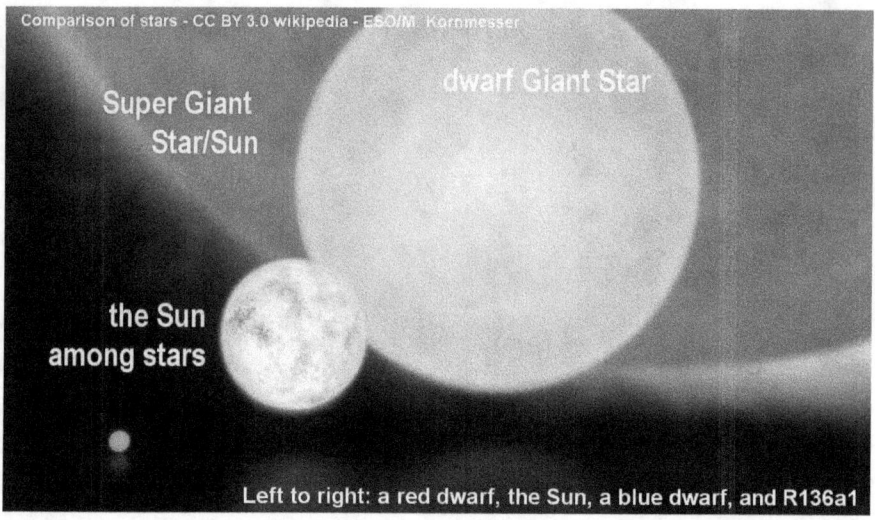

Comparison of stars - CC BY 3.0 wikipedia - ESO/M. Kornmesser

Super Giant
Star/Sun

dwarf Giant Star

the Sun
among stars

Left to right: a red dwarf, the Sun, a blue dwarf, and R136a1

It may also be the case that solar cut-off events may not happen near our space for a few decades yet, till the weakening has sufficiently progressed, and even then, there may not be many candidates for it.

Our Sun is officially designated as a weak "yellow dwarf" which in addition is located in a relatively sparsely populated part of space where it is more vulnerable to weakening supply conditions.

Our Sun, as a G-Class star

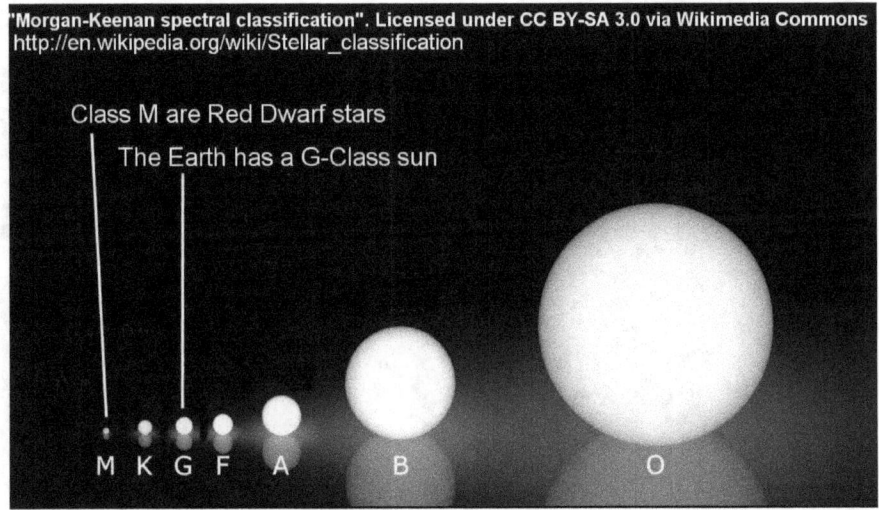

Class M are Red Dwarf stars

The Earth has a G-Class sun

M K G F A B O

Our Sun, as a G-Class star, is not a particularly strong star, though stronger than most.

Smaller and dimmer stars than our Sun

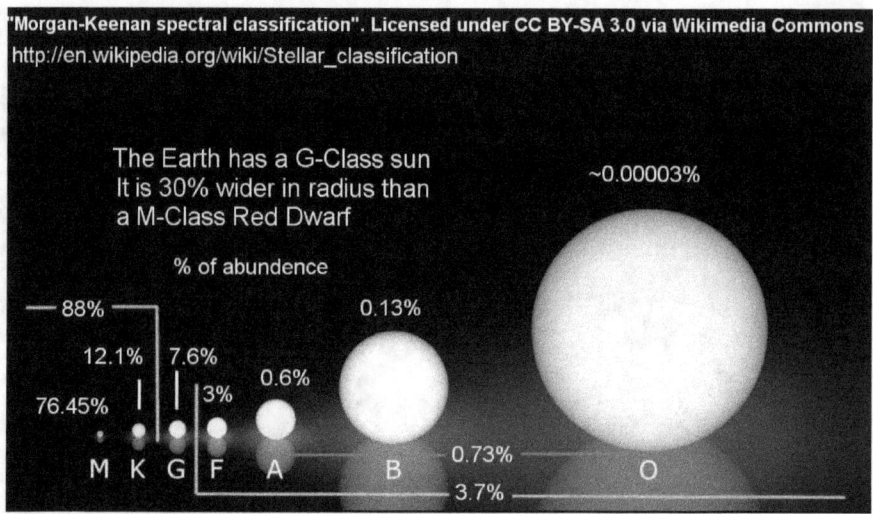

The Earth has a G-Class sun
It is 30% wider in radius than
a M-Class Red Dwarf

~0.00003%

% of abundance

88%

12.1%   7.6%

76.45%   3%

0.6%

0.13%

0.73%

3.7%

M   K   G   F   A   B   O

Up to 88% of the stars in the galaxy are smaller and dimmer stars than our Sun with a radically lower rate of plasma consumption to sustain their lesser fusion reactions. The great giant stars, in comparison are rare, at less than 3/4 of a percent of the stars. The giants are typically located at the center of a large plasma supply system.

The F-Class stars

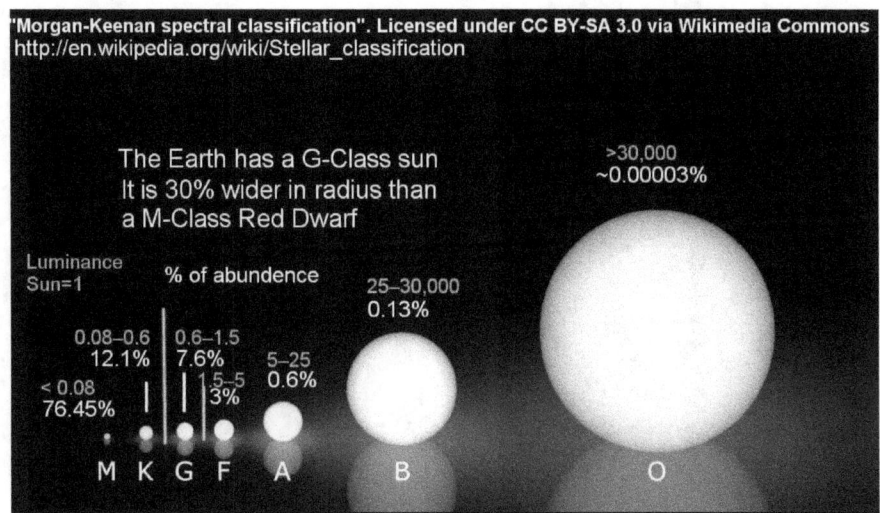

The F-Class stars, that are slightly larger than our Sun and make up 3% of all the stars, are several times more luminous than our Sun, which means that they too, operate in a relatively strong supply system. Those stars may have been the class that was selected for study by the Kepler mission.

All this means that the stars the size of our Sun, which are up to 5 times dimmer, may be the most vulnerable of the stars to supply fluctuations, but might not have been studied by the Kepler mission, because of their lower intensity.

The Kepler spacecraft operates a single instrument

The Kepler spacecraft operates a single instrument, a photometer, which continually monitors the brightness of over 145,000 specifically selected stars in a fixed field of view. The selected stars may have been mostly the larger F-Class stars for their brightness, which would likely be less affected by weakening conditions in the space around their solar systems. All this means that it cannot be said that the Kepler mission has delivered absolute proof that no sun anywhere has ever gone inactive, as has been suggested. When the inactivation of a sun occurs, which should typically be rare and occur quickly, the result would likely be missed unless one specifically looks for such events. The Kepler mission, officially, detected no such events, nor was it designed to do so. It operated for a brief span from December 2009 till May 11, 2013. While it operated, a greater-than-expected noise to data ratio was experienced, from both the stars and the spacecraft itself, which means that the results were not as clean as expected and open to errors.

Inactive stars termed the white dwarf

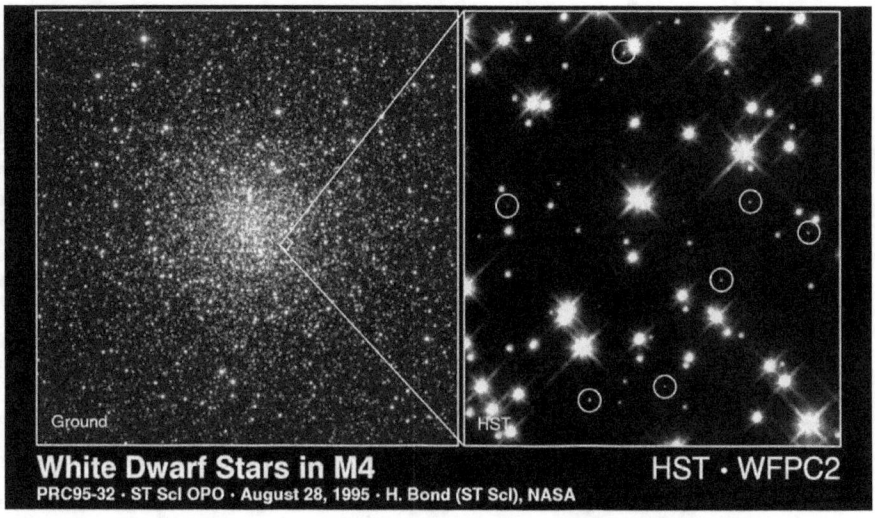

White Dwarf Stars in M4   HST · WFPC2
PRC95-32 · ST ScI OPO · August 28, 1995 · H. Bond (ST ScI), NASA

Some evidence, however, does exist, for the existence of inactive
stars. The visible form of inactive stars is termed the white dwarf.
White dwarf starts are shown here with a circle around them.
When a plasma-powered sun goes inactive, it continues to attract
with its gravity whatever atomic material exists in its surrounding.
The attracted material then filters down to the sun's core where the
atoms become crushed by the still existing large gravity. The
resulting nuclear fission in the core emits the tiny speck of bright
light that is seen at the center of a previous sun. The brilliant light
from the atomic fissioning would be rather small in size in
comparison with the actual size of the then dark plasma sphere that
once was a sun.

Evidence of the inactive state of a Sun

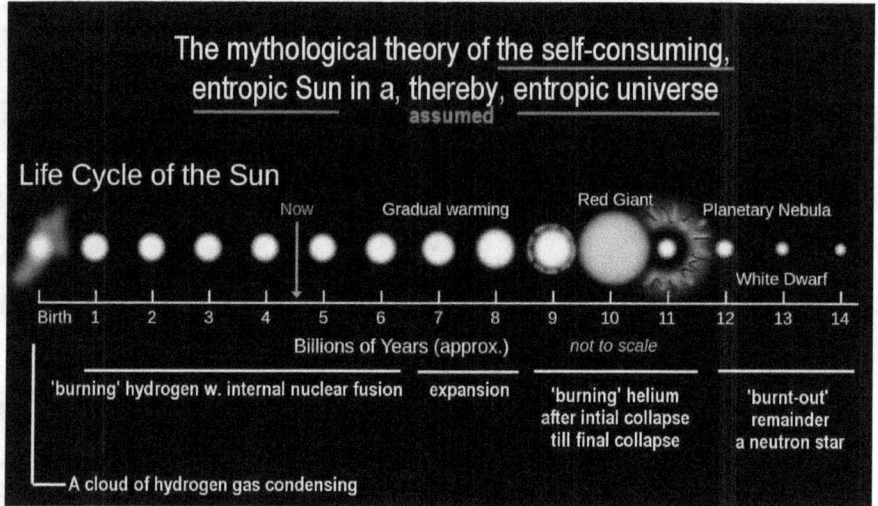

Of course this evidence too, of the inactive state of a Sun, is being sanitized into obscurity by the Big Bang related theory of the entropic universe where every sun is doomed to eventually collapse by explosion or implosion and is left to die away into nothing, with the white dwarf at the end stage.

The fascination with false theories

False Theories

The theory that war is inevitable and beneficial;
...that nuclear war can be avoided with the terror-threats;
...that an economy is benefitted by financial mass-looting;
...that humanity will be saved by depopulation of the Earth;
...that fascism will stabilize the imperial landscape.
...that a limmited nuclear war is possible.

Thus, the global landscape has become a landscape of many obviously false theories. The fascination with false theories has persisted from historic time to ours, such as the widely held theory that war is inevitable and beneficial; or the modern theory that nuclear war can be avoided with the terror-threat of wielding many tens of thousands of nuclear bombs; and also that an economy is benefitted by the processes of financial mass-looting; and that humanity will be saved by the coveted depopulation of the Earth with genocide, as we have it now in progress, and that such a pursuit is a sane objective, so that its fascism will stabilize the imperial landscape.

On the science front

False Theories

Global warming;
The Big Bang creation;
The internally heated hydrogen-fusion Sun;
Neutron stars;
black holes in space;
Stars collapsing into super-nova shockwave fireworks.

On the science front the evidently false theories have likewise many prominent candidates, such as the global warming theory; the theory of the Big Bang creation; the theory of the internally heated hydrogen-fusion Sun; the theories of neutron stars; theories about black holes in space; and about stars collapsing into super-nova shockwave fireworks.

Hypothesis that bring us closer to the truth

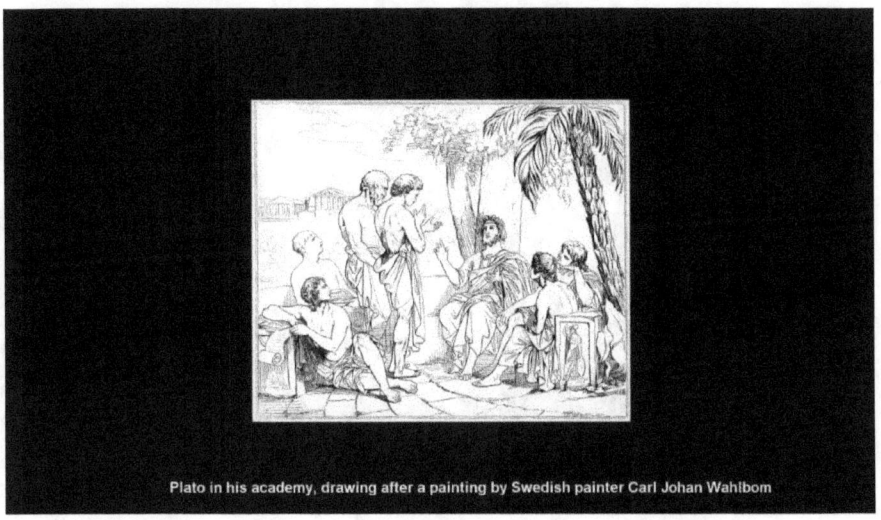

Plato in his academy, drawing after a painting by Swedish painter Carl Johan Wahlbom

Fortunately, as with all false theories, their dominant standing can fade into nothing when discoveries of truth raise perceptions to higher platforms where old hypothesis are simply superseded with higher hypothesis that bring us closer to the truth.

An error in science fades into nothing all by itself

An error in science fades into nothing all by itself when the truth becomes more clearly recognized, understood, and acknowledged. This too is known since ancient days.

We stand at the threshold today

Here we see the path for our future, the path of human development built on scientific development. I like to predict that we stand at the threshold today for a breakout from many false doctrines that have kept us tied to the dust of the earth, the dust of impotence, for far too many ages already, whereby we discover us as human beings with a quality in scientific truthfulness and creative power that renders us collectively as the supreme being on Earth, second to none.

Path of our self-discovery as humanity

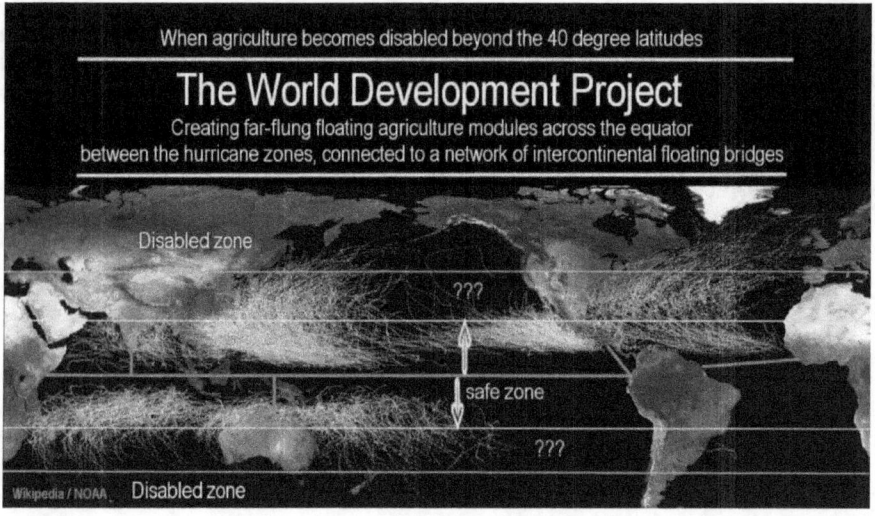

On this path of our self-discovery as humanity, we will meet the immense Ice Age Challenge most surely along the way of it, perhaps even as a secondary issue in the flow of the unfolding power of our universal humanity.